Respectful, modest, enthusiastic, moving: Mündel's "My Life among the Paniyas of the Nilgiri Hills" merits all of these adjectives and more. That the young Canadian agriculturalist could enter so fully into the life of twenty-five Paniya tribal families only recently freed from the status of bonded labourers is nothing short of remarkable. Far from making himself the hero of this anecdote-rich account, Mündel's focus is on the openness of a warm and resilient people whose trust in him overcame the barrier of language and went well beyond agricultural matters. Revisiting the colony with his son in 1986, Mündel noted with satisfaction its new crops and amenities, but he perhaps took a special, quiet, pride in seeing that the timid children he had enrolled in school twenty years earlier had become "some of the real 'movers and shakers'" in their community.

CUSO alumni from the 1960s frequently observe that they got back far more than they gave during their brief assignments in the developing world. Probably so: they were young and naïve and inexperienced. But they were also enthusiastic. And when that quality was wedded to common sense and compassion and a stick-to-it attitude, as in Mündel's case, the giving flowed both ways.

Ruth Compton Brouwer,
Professor and Chair of History at King's University College,
University of Western Ontario, London, ON.
Has published two books and many articles on Canadian women in colonial Asia and Africa. Her current research deals with CUSO volunteers in the 1960s and 1970s.

The time is 1966, the place, a remote part of Tamilnadu in South India near to its border of Kerala. Hans-Henning a young German from Canada who has been studying agriculture has been sent on a volunteer assignment to India by CUSO, a Canadian equivalent of VSO. This placement has been organised at the request of Dr Narasimhan, a champion of the Adivasis or tribal groups living in the Nilgiri Hills....

Dr Narasimhan was one of those rare souls who not only takes the trouble to comprehend how a wrong has come about but actually takes action to put it to rights. .... And so it was that Hans-Henning, the author of this book, as a 24-year-old agricultural student from Canada found himself appointed Farm Manager for a group of illiterate and poverty stricken families with the objective of helping them to become a cooperative agricultural colony.

This book is Henning's fascinating account of the months he spent there, put together forty years later from the letters he wrote at the time to his fiancée and parents in Canada. ...

His account brings to light many obstacles and difficulties he had to face but it also shows his remarkable resilience and open-mindedness and considerable success in making friends with the families in his charge, not to mention what was achieved in the planting and husbanding of a variety of crops.....

Hans-Henning deserves praise for the efforts he put into a brave experiment as well as for this vivid chronicle of it.

Rosaleen Mulji
an ex-officio member of the Managing Committee of the Nilgiris
Adivasi Welfare Association in Kotagiri
Secretary of the Nilgiris Adivasi Trust in England

My Life Among the Paniyas of the Nilgiri Hills

With an Introduction
by Clyde Sanger

Hans-Henning Mündel

My Life Among the Paniyas of the Nilgiri Hills

ISBN 978-0-9783593-0-0

Carpe Diem Mündel Publishing
P.O. Box 21031
West View Mall
Lethbridge, Alberta
T1K 6X4

e-mail: hhmuend@shockware.com

Dedicated to:

Bev, my wife (then fiancée)

and my mother, Gisela Mündel ("Mutti"),

and in memory of my father, Hans-Erich Mündel

This book is based on letters to them.

Acknowledgements

First of all, I want to thank my wife, Bev, and my parents, Hans-Erich and Gisela Mündel, for saving, and then subsequently passing on to me some years ago, all the letters upon which this book is based. Had they not saved all my letters, sent to them during my time at the Paniya Colony in 1966 and 1967, this book could never have been realized. Bev really wanted this information to be made available in a readable form, especially for our children and their families. Thus the idea of writing a book about my life among the Paniyas took form. The task, at times, though, proved daunting. I had written all the letters in a very tiny script in order to cram as much information as possible onto a page, as stamps at the time were very expensive in India. As a result, deciphering the letters became quite a chore. Also, as Bev and I were engaged at the time, too much personal information and items related to her life in California, which were not really relevant to the topic of this "diary", were included in those letters. Therefore they had to be carefully gone through and mainly references to my stay at the Colony excerpted for this book.

On March 31, 2007, my 65th birthday, I retired and was ready to start on this project. Within two weeks, on April 14, Kathleen Mailer made a brief presentation on her book, How to Write and Publish Your OWN Book, at the Chapters bookstore in Lethbridge. Bev and I attended. Kathleen and I were able to have a bit of a chat and at her suggestion, I sent her some pages of my "diary" from my time among the Paniyas. She greatly encouraged me to proceed and quickly enlisted for the project her graphic designer, Dominique (Nik) Petersen, who promptly developed a draft front and back cover. I was now spurred on to get the book written. Thanks so much for all your encouragement, Kathleen! And thanks, Nik, for your patience in working with me, for the cover and subsequently formatting the book. This also included ensuring that each of the 40 pictures was suitably sized. Most sincere thanks also to Jens Petersen for his able editing. Throughout the editing, Jens bore in mind my initial concern for maintaining the authenticity of the times (the 1960s) in the letters. I also wanted to ensure that my personality, my observations, my enthusiasm, yes, my passion, as well as my frustration and indeed anguish still came through in the writing. Thanks, Jens, I think we managed that! And Bev, your constructive final complete read and comments, just prior to the book formatting, were invaluable! Thanks!

In June 2007, I contacted the Nilgiris Adivasi Welfare Association to apprise them of my plan to write this book, sending them the draft front and back cover. The encouragement of K. Vijayakumar, in charge of the NAWA administration for most of the past decade, and whom I had met on one of my 1980s visits, gave a real boost to my writing of this book. It is from him I then received the request for fundraising for the tribal hospital at the Paniya Colony to help the tribal peoples of that region. Those readers moved to help the Paniyas further may please contribute directly to NAWA or NAT (details provided at the end of this book). You should also know that a fixed amount of the purchase price of this book will go towards that hospital project.

Thank you to those of you who so kindly read and commented on the whole manuscript! While your names appear elsewhere, your efforts are hereby thankfully recognized.

And finally to Clyde Sanger, a special thank you for agreeing to write a Foreword, without even having read my manuscript! I have known Clyde since the early 1980s and have, over the decades, really appreciated his writings on international development.

20 October 2007

Table of Selected Topics

1967

Picture Captions

Foreword

This question is for you, dear Reader. Did you miss out, in your idealistic twenties, on a memorable adventure that might have severely tested your character and yet probably turned out to be the most exhilarating time of your life? Maybe you did. Lucky those few thousand in the richer countries of the world who came to adulthood in the 1960s, who had completed a degree or two and who were caught up in the wave of secular (well, in some cases missionary) zeal to help folks far away who were leading fairly miserable lives. Most of them, Henning among them, seem to have had great experiences.

The pebble that Alec Dickson dropped in the ocean in 1958 when he launched Voluntary Service Overseas in Britain produced a ripple across the Atlantic. Two years later Senator Hubert Humphrey and the Kennedy brothers launched the Peace Corps. In our own country the Canadian University Service Overseas (now CUSO) grew out of the initiatives of a few young people who had found an echo from a single expedition - Lewis Perinbam in the famous 1957 WUSC study trip (it included Pierre Trudeau) to Ghana, Guy Arnold and his school leavers in Guyana, and Keith Spicer with his 16 volunteers who went to South Asia in June 1961 under the name Canadian Overseas Volunteers.

By the time Hans-Henning Mündel joined CUSO in 1966 and went to India for three years, the organization had matured and expanded: in that year it had 560 volunteers placed in 34 countries. Henning was among 32 CUSO in India, and he was almost certainly more qualified than the pioneers. He had a BSc in Agriculture from the University of British Columbia, and two Masters degrees – one in agronomy and plant breeding, the other in international agricultural development – from the University of California, Davis. He had also seen, from the age of three, more of the tougher side of life than most Canadians. Fleeing in 1945 from Russian army forces to refugee quarters in Düsseldorf, his family who were originally German Balts from Latvia only reached British Columbia and the orchard country of the Okanagan six years later.

So he was a well traveled, well schooled agriculturalist when he was posted to India, to work with tribal families in the Nilgiri Hills in Madras State (now Tamil Nadu) as farm manager of the Paniya Colony. The Madras government had just granted to the Nilgiris Adivasi Welfare Association (NAWA) 100 acres of land, for 25 Paniya families to work. He was only 24 and the farm was very isolated. His living quarters were extremely primitive: he occupied the smaller of two rooms that together measured 97 square feet and had floors made of cow dung and mud. The larger room was a kitchen where Anthony, a charming older man whom NAWA had hired as his 'butler', stayed. There was then only one well on the farm, and washing facilities consisted of a jug and basin on a stump.

Daunting enough. Moreover, the Paniyas were illiterate and (as Henning later wrote in Man Deserves Man, the CUSO anthology published in 1968) "reputed to be perhaps the most backward of the many Nilgiri Hills' tribes – despised by much of the local population for their filthiness and poverty, landless labourers, formerly working as bonded servants, almost like slaves". One can understand why David Hopper, who was agricultural adviser to CUSO-India on top of his job with the Rockefeller Foundation (and who in 1970 became president of the International Development Research Centre) tried to change Henning's placement as soon as he landed in Delhi. Hopper sent young Henning off to see a plant breeder who was doing agronomic research on a range of crops at Phaltan in Maharasthra state, and it was assumed he would stay there. Henning's spirit comes through in a short sentence: "However, I was on my way to the Nilgiri Hills and, as you can read in this book, I stayed there for 10 months."

So this is a book about those 10 months. It reads like a diary, but is an edited collection of the letters he wrote to Bev Atherstone, the fiancée he had left completing her psychology studies in California and (in German) to his parents. They were months of hardship and joy; of facing the language barrier and strange customs with gentleness, clear affection and good humour; of making considerable advances to diversify from a staple crop of tapioca (or cassava). Under his direction, the Paniyas grew ginger, lemongrass for its oil and planted mulberry bushes for silkworm culture, as well as the young fruit trees, mangoes and limes. The rains, though, were unreliable and in March the severe drought killed the beans, chickens ate the cabbage and tomato plants, and elephants smashed banana trees nearby.

Through all this, the men showed their sense of humour. One day, he came on them planting tapioca in an odd pattern, stretching some 40 feet. When he looked carefully and asked the supervisor, he realized it was a caricature of himself, standing with arms akimbo, designed by Veran a part-time hypnotist. In turn, he taught them to plant along the contours, insisted on an iron plough in place of their homemade wooden ones, and purchased a bullock cart that could bring in manure.

And affection. He persuaded the local elementary school to take 20 Paniya children on condition they were scrubbed and properly clothed. Henning and Anthony set to soaping them at the well, and later sewing on buttons and mending torn clothes. He oiled and combed their hair and used to walk them to school. After a while he had to spread his fingers wide as so many of them wanted to hold his hand. The CUSO staff in Delhi took to calling him "Mother Hen". And the children put roses in front of a photograph of Bev.

Isolation was a huge factor. It was 120 miles to NAWA headquarters and his mentor Dr. S. Narasimhan, who visited at most once a week, and five miles over the hills to a tea estate in an emergency for other doctors. He provided pills and first aid. Many Paniyas suffered from anaemia, workers cut their feet badly and dear Anthony died of dysentery. It was a nine mile round-trip by foot to a post office and it took three months for parcels to arrive from Canada. But Henning from his letters (and of course you pump yourself up to be cheerful in such circumstances) hardly seemed to have had a low moment, except when the fertilizer didn't come nor the man to take soil samples. He welcomed the white ants invading his small room, the rain deluging through the straw roof, and he gloried in the evening spectacle of thousands of fireflies and marveled at the mountains. He once acted as a midwife, gave vitamin pills to nursing mothers and organized midday meals. He played his recorder, took lessons in Malayalam from the children and wrote full descriptions of weddings and funerals. And he seemed to love walking those many, many miles. "It is one of the most beautiful places in the world," he wrote.

But money was always short. His monthly salary from CUSO and NAWA combined was 85 rupees – about C$11.50. (German volunteers got 600 rupees.) In April a letter from NAWA announced strict austerity and cut the midday meals. And a letter from Bev asking "whether I felt I was actually doing something real here" was an obvious jolt into introspection. He wrote that his contribution was "vastly suboptimal" as his academic qualifications were inadequate and wrongly slanted for this place, he hadn't mastered the language and generally lacked practical experience. He decided of his own accord that he was a financial weight on the farm and should leave the Paniya Colony. David Hopper thus won his point in the end, as Henning transferred in August 1967 to the Phaltan research institute.

So to my personal view on his book. In researching the book that I wrote in 1969 (*Half a Loaf: Canada's semi-role in developing countries*) I visited CUSO volunteers in Jamaica, Colombia, Nigeria and India. In three subsequent years when I was on the CUSO board of directors I visited others in Tanzania, Malaysia, Thailand and Papua New Guinea, and have kept up an association with many returned cooperants, as they now call themselves. And of course I have read the CUSO anthology Man Deserves Man to which Henning and 28 others contributed, as well as Ian Smillie's history of CUSO, The Land of Lost Content. So it is on this basis that I really think Henning's book ranks with the best writing on this fascinating organization and the testing life of its volunteers and is a wondrous window into those early days. As a collection of letters, it is better than journals and diaries that too often are self-serving. It is lyrical, but you are also in the mud of those distant fields. It is shiningly genuine, because you can't fool your fiancée or your parents. It has been the most enjoyable read I have had for months. I hope you relish it as much.

Clyde Sanger, October 2007.

Preface

PANIYAS AND NAWA

For many years the Paniyas have lived as bonded servants to smaller landowners in the region of southwestern India. As the anthropologist, Edgar Thurston, stated in 1909: "Their position is said to be very little removed from that of a slave, for every Paniyan is some landlord's 'man', and though he is, of course, free to leave his master, he is at once traced, and good care is taken that he does not get employment elsewhere." [1]

Dr. S. Narasimhan (died 1978) was a Tamil Indian medical doctor, a Brahmin Hindu, born and raised in Ootacamund (Ooty), who had devoted his practice to the welfare of the tribal people of the Nilgiri Hills, beginning in 1942. His paying practice helped him to provide such services free of charge to the poorest of the poor. Eventually, in 1958, at the insistence of close personal friends, like the retired Parsee civil servant from Bombay, Mr. Phiroze Ghandy, an association was established, the Nilgiris Adivasi Welfare Association (NAWA), which enabled the raising of funds to carry on and expand this work. "Adivasi" is a general term for tribal peoples of India; it may literally mean "those who lived here before" or "original inhabitants". I think it is out of a combination of both respect and affection that Dr. Narasimhan was generally known within NAWA and beyond, simply

[1] Thurston, Edgar. 1909 (reprinted 1975 by Cosmo Publications, Delhi). "Paniyan." In *Castes and Tribes of Southern India*, edited by Edgar Thurston and Kadamki Rangachari. Vol. 6, 57-71. Madras: Government Press.

as "Doctor". And it is as "Doctor" that he was introduced to me when we first met in September 1966.

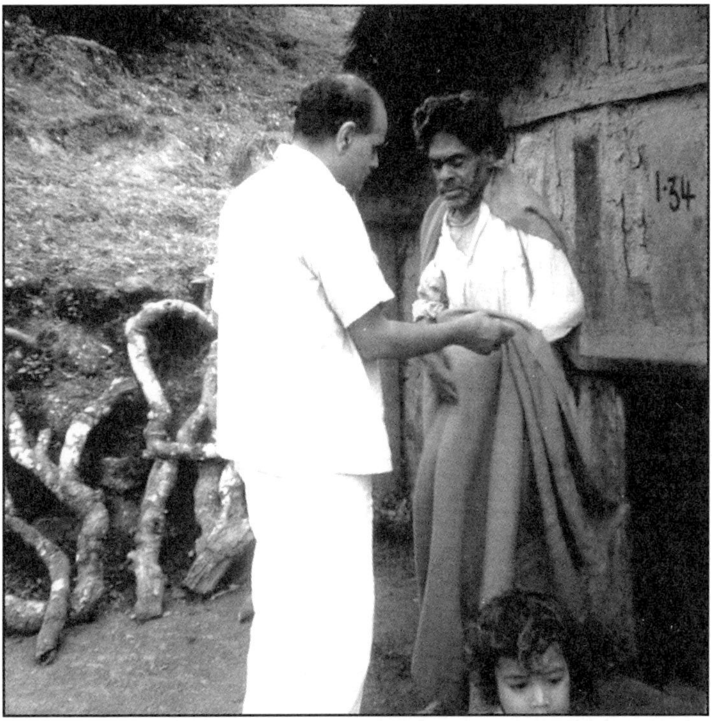

Dr. Narasimhan treating a Toda patient

Two "western" ladies joined NAWA in the months before I arrived there. Miss Victoria Armstrong (died in India, 1997), an inspectress of schools in Leeds, England, took early retirement to devote her full-time and considerable energy to NAWA, becoming, as its "Honorary Secretary", its fundraiser par excellence, globally. Miss Martin (died in India, 1972), a bookseller from Australia, selling Australian text books to schools in India, volunteered as Treasurer of NAWA. Frequently these two ladies would accompany the Doctor on his nighttime visits to the various jungle hospitals operated by NAWA.

NAWA and Dr. Narasimhan's headquarters were at Kotagiri, about 30 miles north of Ootacamund (Ooty), the headquarters town of the Nilgiris District.

A recent description of NAWA is as follows: "NAWA is a tribal welfare organisation working for the welfare and development of the scattered aboriginal tribes of Nilgiris in Tamil Nadu and adjoining areas covering a tribal population of 30,000; [it] runs hospitals, dispensaries, crèches; [it] conducts education and economic development programmes and rehabilitation programmes for bonded labourers."[2]

PANIYA COLONY

Dr. Narasimhan heard of patients requiring his help on his trips into the Paniya region. These shy people, however, were reticent to approach him, a non-Paniyan, for help. As his jeep approached, the

Chunda, Paniya woman

[2] http://www.wiserearth.org/organization/view/7bb34a030a855cd02ffe91df6ecf6dcc/

Paniyas would run away. However, a courageous young woman, Chunda, persuaded her fellow-Paniyas to trust the doctor.

And thus began an association with Dr. Narasimhan and NAWA. The landless plight of the Paniyas motivated Dr. Narasimhan to seek help. Just a few months before I joined the Paniya Colony, negotiations with the Madras State Government and NAWA had been completed with a grant of 100 acres of land for this Colony. This was just a few miles from the border to Kerala, on the western slopes of the Nilgiri Hills, a bit over 120 miles from NAWA headquarters in Kotagiri. In 1898, on these western slopes of the Nilgiri Hills (as told to me by Dr. Narasimhan), Ronald Ross, accompanied by his grandfather, identified the *Anopheles* mosquito as being responsible for transmitting the malaria parasite to humans.[3] A subsidy/loan was provided to NAWA and five long houses with five family units each, consisting of two rooms, were built by the Welfare Department. A grant towards building a community kitchen was also provided. A well for drinking water and washing was also constructed for the Colony. And then the 25 families, 120 people, of the Paniya tribe were moved into these houses. Some of the Paniyas had been living on that land already, the Karikkamunda group. Thus we called this Colony the Paniya Colony Karikkamunda. The oldest male member of this group, Vella, henceforth considered himself the "chief" among this group of Paniyas. Chunda was Vella's younger sister.

Dr. Narasimhan knew the Colony could not be managed by long distance, 120 miles from his headquarters at Kotagiri, 90 miles from the Nilgiris District Headquarters, Ooty. CUSO had contacted him and he knew what he wanted – a Farm Manager. In October 1966 I became that Farm Manager. And my experience among the Paniyas for the following 10 months is the essence of the following book.

The long-term goal has been to turn the land over to the Paniyas themselves. The aim of NAWA (including me) for the Paniya Colony was to diversify the agriculture, thus to minimize loss should a certain

[3] Kapur, A.P. 1958. A report reviewing entomological problems in the humid tropical regions of South Asia. in Anon. (Ed.) *Problems of Humid Tropical Regions, Humid Tropics Research*, UNESCO, Paris. pp 59-72.

crop fail in any given season. We started in 1966 with the planting of tapioca. The starchy root of this plant, also known as cassava or manioc, is a staple diet of the Paniyas. We also grew lemon grass, the oil of which we extracted in a little still on the Colony and sold into the market. We planted mulberries for eventual silkworm rearing and also mangoes and limes and some vegetables, as well as ginger.

I had just completed two Masters degrees at the University of California (Davis) (UCD). One MS was in agronomy/plant breeding and the other was in international agricultural development. With my interest in international agricultural development having evolved over some years, I wanted to obtain some practical first-hand experience in the "developing world" prior to continuing for a PhD. Thus I approached CUSO in Ottawa. I was asked to the UC-Berkeley campus to meet for an interview. This took place over coffee, outside on the university lawns, with a Canadian lady professor, who had been involved in the formative years of CUSO (which started in 1961). I was accepted and in July began two months of intensive general orientation at UBC in Vancouver, BC, Canada, for deployment to a "developing country", specifically India. The morning-to-night curriculum included culture, customs, and the languages of southern India. During this time, CUSO had arranged for me to take up the position of "Farm Manager" with the Paniya Colony being operated by NAWA. However, as soon as I got off the plane in New Delhi, Dr. David Hopper of the Rockefeller Foundation and CUSO-India agricultural advisor, took me aside and indicated that with my Masters in agriculture, he had a more appropriate placement for me in mind. This was with a seed producer in Maharashtra State, who was also involved in plant breeding and agronomic research on various crops. While I appreciated Dr. Hopper's concern, I felt I needed to go to the Nilgiris and see what NAWA had in place for looking after the Paniya Colony. After all, CUSO had made the commitment that I would come there, to NAWA. Nevertheless, Dr. Hopper arranged for me to meet Dr. Lee House, a colleague of his, in Bombay, and we travelled together to Phaltan, in Satara District of Maharashtra State, where I met Mr. B. Nimbkar and his team. Mr. Nimbkar assumed I would stay right there with him. However, I was on my way to the Nilgiri Hills, and as you can read in this book, I stayed there until the following July.

What follows is in the form of a diary. I based this entirely on two sets of letters. One group I wrote to my then-fiancée, now wife, Bev in California, USA, where she was continuing her studies at Davis. The other group of letters, all written in German, was to my parents, in Oliver, BC, Canada. The words and thoughts are mine at age twenty-four, and the times were the 1960s.

Hans-Henning Mündel, 18 June 2007

 Map of Nilgiris District and India

1966

INTRODUCTION TO CUSO ASSIGNMENT AND DR. NARASIMHAN'S REALM

2 October 1966, Sunday

Kotagiri

Planning to "help India" is definitely too grand and presumptuous a notion for me. Although I was asked about this before leaving Canada, it is not at all my aim in coming here. If I can, however, help these 25 Paniya families (especially if I can attain a certain fluency in Malayalam), I will feel that I have contributed something real and useful. There is so much that needs to be done immediately – and later – and eventually. And so much is to be learned about these peoples' ways of thinking, their sensitivities, etc. This is where having an interpreter who speaks very poor English will not be of much help. I must learn the language, and I hope the day-to-day chores will leave me enough time for language study! How can the Paniyas be convinced, although now paid for their labour, that the land will soon be theirs and so will the crops and the financial returns? Not an easy problem!

Dr. Narasimhan is really a person to be admired. He does get very little sleep! I liked his statement yesterday, that he doesn't want the Paniyas to work seven days a week, just for the sake of some extra money, but that they should have Sunday as a day of rest. The Doctor is a Brahmin. Apparently he is quite sincere and strict, but without many of the so-often-associated rituals. The Doctor's medical services to all the hill tribes here in the Nilgiris are free of charge.

Commenting on whether we should try planting some apples in our Paniya Colony at Erumad, he stated that he is the apples' rival. Apples are expensive; his services free. Thus: "the Doctor a day keeps the apples away!". I thought that this was a rather cute twist! We will go down to Erumad tomorrow. If all the preparations in the hut are completed, I'll return there with my belongings in the course of the next few days.

I'm also spending some time reading up on diverse crops. I went to Coonoor to find out what plants are available. Not much now! This is a land where mountains are called hills; houses, bungalows; rains, monsoons; and forest, jungle.

9 October 1966, Sunday

Kotagiri

The Doctor is approached by people from all over the place in connection with the hill tribes: from Delhi, from abroad, from ministries, and by anthropologists, etc. And many tribals approach him with all kinds of problems, requesting him to take them up on their behalf with the appropriate authorities. This he does to a considerable extent. In relation to a Toda's petition given him, he told me that once these Todas roamed all over these hills, and now they have to petition to get one acre of land to live off!

The other day in the Clinic, the Doctor introduced me to a lady of the Badaga tribe. The Badagas are the most de-tribalized of the Nilgiri Hill tribes, and most akin to the Hindus. She was the first woman university graduate of the Badagas, and now is a Member in the national Parliament, representing the whole Nilgiris District. The largest percentage of the NAWA membership is from the ranks of the Badagas. Most of them started out in rather severe privation and are now the first generation above chronic poverty.

SETTLING IN AT THE PANIYA COLONY

10 October 1966, Monday

Near Erumad (at the Paniya Colony Karikkamunda)

On a tropical night with its tropical sounds (some very loud cricket-type creatures – absolutely deafening), I'm sitting by the light of my kerosene lamp on the verandah of one of those huts which just a month ago I first saw and thought so poetical and picturesque.

The picturesque part absolutely can't be denied. But the poetical part really is debatable! – at least until a closet and cupboard are eventually

Henning in front of mud hut home

built by the carpenter; and until my butler, Anthony, moves from the other room into the kitchen. It is somewhat prosaic to be cramped into a tiny room 6.5 x 7 ft. – i.e. 45.5 square feet. The other room, which is now occupied by Anthony and cooking utensils, is 6.5 x 8 ft. – i.e. 52 square feet. I think Anthony would prefer to stay there. But I must have a place to hang my clothes up. They can't be cramped in a suitcase for two years! And I need to be able to lock up mine and the Association's money for wages. All told, my effective room space will be a floor area of less than 100 ft. square, consisting of a mixture of cow dung and mud.

A few more days and the present "tea house", which Kurian, the overseer, uses this hut for, will be relocated. Right now Kurian uses this hut for serving tea and snacks to the Paniyas and road workers in the morning and evening. A thatched roof, supported by posts, is being built a little way down the road to serve as "tea house".

At present I'm a little, to say the least, upset. Perhaps later I'll again be able to see more of the poetical! Why am I upset? Well, first of all, my two days here were rather exhausting, just walking up and down and up and down the hill slopes of our 100 acres. This Colony farm sure seems large to me! I'm really out of shape! I perspire constantly, and it's rather hot. And I have quite a sunburn too. I am adopting the "neat" custom here of using an umbrella as both parasol against the sun, and umbrella against the rain.

The Paniyas seem to be really very sturdy, hard workers, even though they did quit yesterday because of a little rain. That was just before the heavy rain came down, and the mulberries needed to be planted. The Paniyas left the young plants in the pits, without filling them up with soil.

Kurian is the owner of the hut I'm staying in. He really knows very little English. Thus Anthony is now my life-saver! – as interpreter, and the only person I can actually talk to relatively intelligently. Mind you, I still need to use simple words to be understood. Also he cooks very nicely and seems quite concerned about my welfare. Mainly he cooks "western-style" food, called "European food" here. I am trying to encourage Anthony to also cook some Indian food and use some more spices. We'll see how that will work! NAWA has hired Anthony for me. He also goes grocery shopping for me and washes my clothes by beating them on a large stone, using hot water and soap. Then he irons the clothes with an iron filled with glowing pieces of charcoal. Without Anthony, and having neither fridge nor electricity, I would have to spend almost the whole day just looking after myself. The nearest groceries are available two and a half to three miles away, by taking a shortcut through the jungle to the small village of Cherambadi.[4] Rancid butter, I've found out, can still be used in place of cooking fat.

Mulberries – we are going into the silk business, but probably not till after I leave. Here's an example of how "un-slow" Indians move. I had visited the federal sericulturist (silk worm specialist) in Coonoor about 10 days ago and merely enquired about the possibility of getting planting material. On the day we came here, the local government sericulture manager, responsible for a 32-acre silk farm at Erumad, had been looking for the Canadian to deliver the plants to, sent from near Bangalore (in Mysore State). I wasn't even here yet! The evening I moved here the plants were in our possession – free, since the government wants to encourage silk farming. The "jungle telegraph" works fast here.

The glow-worms are just so wonderful in their abundance at night! With the aid of a flashlight, I picked one up and saw that the light comes from two tiny spots on the lower side of the "worm's" tail. And during the day mynah birds are abundant. As I walk along, I can't resist now and then brushing the low-creeping mimosa (touch-me-nots) on purpose, to watch a whole branch or plant wither, but only temporarily. Some psychologists might construe this action of mine as a need to feel power over something or other. Actually though, I just find it so intriguing. Perhaps I'm childlike in this respect! For entomologists this would really be heaven. I have already seen many interesting and even beautiful bugs and beetles and many, many beautiful large blue butterflies with white or yellow dots. I've also seen a huge butterfly with a deep red body, framed in black and having black wings with white speckles. Also quite large and neatly

[4] Cherambadi is located at 11° 32' North, 76° 17' East

striped lizards abound. Someone showed me local porcupine needles, up to one and a half feet long and some about pencil thick! I hear many bird songs and voices which are still quite foreign to me. Some are quite beautiful songs, others interesting, too. The flora has abundant blooms, some delicate, others substantial, yet I hardly know any of their names!

Last night, where my "washroom" is to be, I had the first water to wash in since I left Kotagiri the day before. We had to find a pail and arrange for someone to regularly get water from the well. This is settled now. Today sacks have been put up around the sides of my washing facility. I am under the roof of the hut, but have no walls. A wash basin on a stump serves as my sink. A jug filled with water stands beside it, and I use a small-handled cup to pour water over myself, as a shower. At times, the cow tethered to a corner post of the hut nearly reaches my wash basin. In spite of the fact that she really tries hard to disturb me, I don't want to disturb her.

Fortunately, a lot of the work happens as though by itself. This gives me an opportunity to visit the different work groups in the mornings and afternoons, checking who is there, and who does what. In this way I hope to establish how long different jobs take and what can be expected from the different men. As I don't know our Colony boundaries very well yet, I have Kurian show these to me. In one area I saw the jungle literally fighting back. So I hastened to assign a weeding crew to that area.

I pay the Paniyas every evening because they don't plan ahead too much and would require advance payments if I didn't. I have a receipt book, with one line per Paniyan. It's all purplish from the stamp pad ink of the finger–, no, thumbprints.

We'll see how my mail delivery will work! About every two days I plan to send Anthony the three-odd miles to the post office to check for mail. This post office is only open one hour in the morning and one hour in the evening, and they apparently don't deliver mail.

Tomorrow I must go to Cherambadi, a village two and a half to three miles by shortcut over hill and dale, to get five-Rs notes (rupees)

Anthony and wage receipt book: each line is one Paniya's thumbprint

changed to one-Re's (rupee), for wages! The standard daily wage is two Rs. Since all my travel is on foot, one of these days (or months), I'm going to fire off a letter to those CUSO people in Ottawa, who stress that CUSO discourages volunteers from having servants, to explain that I'd spend all day just going to the villages to do my errands and shopping. With no electricity and no fridge, I can't keep many things more than a few days! And then cooking needs to be done. I would have no time for my job, which isn't a small one! My mouth constantly goes dry, and water (boiled and lukewarm) may be one half-hour away, at my hut.

14 October 1966: Friday

It's so ridiculously hilarious, the situation I find myself in! – a tropical thunderstorm, after a few days without rain. The rice-straw thatch of my hut naturally was dry and shrunken – so the roof was no help! While outside it was pouring, inside it was raining. In about 20 minutes the straw had time to swell, and the rain inside stopped! My shirt is drenched as is everything in my room, except those things in the trunks. As quickly as I could, I got out the plastic garbage bags I had bought in Vancouver and stuffed things into them or spread them overtop. Thanks to my father's suggestion, I had bought a large piece of plastic and could eventually cover everything in my room with it. But I had to get it out of my trunk first. This plastic sheet has been worth its expense already, even if I never need it again. I think I managed to keep most things from being thoroughly drenched, except, of course, for the shirt I have on, since I was constantly bending down. So now I won't have to ask one of the many questions I had in my mind: "What happens when a real rainstorm comes?" Naturally, things get wet! I really don't know why I didn't think of that! I'm really so dense! Although the rain has been pouring for over an hour, with increasing velocity, most parts of the roof are resisting the onslaught of the heaven-to-earth-bound masses of water quite well. Actually, I'm quite proud of my first real tropical thunder- and rainstorm and all that it entails! What false pride! It sure isn't any achievement of mine! Nonetheless, it's thrilling! Of course it's different from camping, because here, who can tell when it will happen again? Anyhow, I do have the plastic and, this is somewhat unseasonable, since the strong rains here generally come with the Southwest Monsoons from July to September; furthermore, in about a month's time, perhaps no more rain will come till next May or June! All day I have been wandering about looking at the clouds, hoping and praying that this evening, the usual time for rain, it would rain. For the past three days there has been no rain whatsoever, and we had planted the mulberries, which desperately need water. I had already planned, in a really half-hearted sort of way and most reluctantly, to send a crew tomorrow to water the mulberries by bucket from the well. The well is very far from the planting and we don't have many buckets! I feel all this as sort of a prelude to experiencing what it means to depend on the weather, to be at the mercy of the elements, as the peasant farmer here is so utterly dependant!

Since I had just finished paying the Paniyas, who again quit at 4:00 p.m., I had many guests on my veranda, or rather said, much companionship in getting wet! Their own houses, built by the government and with tile roofs, would have kept them dry. Slow rain started at 4:15 p.m. and then the elements let loose at 4:30 p.m., as though a major water tap had been turned on! It was quite fascinating, seeing huge drops hitting the ground a couple of hundred feet away but still being completely dry! But that was short-lived!
I'm sitting now on my veranda with a lit kerosene lamp, as it's just after 6 p.m. and getting quite dark. The last Paniyas have just left after enough "Kuttas" (special umbrellas for working under, made of split bamboo and banana leaves), were found to see them home.

Paniya homes three 5-unit (buildings) and Community Kitchen

A messenger who works for Kurian and also sleeps in the back of this hut has now been officially authorized by me to get my mail, since he goes near the post office almost every morning.

Paniyas making "kuttas" (working umbrellas)

In spite of my apparent trials and tribulations, why should I complain? I'm living like a king in comparison to all those around me here. In a way, it's frightful to be greeted with so much respect by so many people! A few mornings ago, when a crew of the Paniyas was having tea here on the bench on the veranda of this "tea house", I came out of my room, and almost all of them stood up and then sat down again. As I walk along the road, so many of the road crew, who chop up the rocks with hammers, greet me in a semi-salute which seems to be rather widespread here. Oh yes, soon this road, the Silver

Hut I lived in for eight months

Jubilee Road[5], will be transformed into a "metalled road", with small chips of rocks compacted into the surface, so that a bus service to Erumad can be started along here. The road is mainly clayey soil now, and this becomes very mushy in rainy weather.

Who knows when my washroom and closet will be built! Finally – yesterday – the latrine was completed. Frankly and honestly, a lousy job – hopefully at least it will be serviceable until the first strong wind hits.

Last night, with the help of Kurian as interpreter, I took a census of our Paniya colony, at the Doctor's request. We reached a total of 117, one of whom was a child born just five days ago. The Doctor hasn't seen the baby yet. Kurian, I suspect, simply estimated most ages. For children who weren't around, he asked the parents to indicate their heights. It was hilarious! Often he just put the wife down as five years younger than her husband. One funny notation was a husband aged 30, his wife aged 25, and their son aged 18. The poor mother would

[5] Probably named in honour of the 25-year reign of King George V of the United Kingdom, in 1935. George was the first Emperor of India to actually visit India, which he did in 1911-1912. He was accompanied by his wife, Queen Mary.

have conceived at 6 from a husband of 11. Obviously something got lost in this translation. But Kurian tried to be so helpful! Other assistants were two Paniyas, Chunda, whom the "Western" ladies call "Cleopatra", and her oldest brother, Vella, the self-styled Paniya leader. Both of them nodded most vigorously when I managed to repeat a name for a son or daughter correctly. The whole thing was so much fun and really a panic. As well as ages, names, and relationships, I had to find out who, and how many, lived in each of the living quarters.

Chunda and another Panichi pounding and winnowing grain

My full postal address is quite intriguing: Nilgiris is the District; Gudalur is the "taluk" (a subdivision of the district); Cherambadi is a larger village where mail first arrives and gets stamped before going on to the Konnachal Post Office, on the outskirts of Erumad, the same morning.

I'm trying to ascertain how much tapioca has been planted so far. Not an easy task! First of all, most rows aren't too straight. Then trees are in the way, with subdivided rows, etc. etc. Every now and then, the Paniyas like to joke too: they will suddenly put an extra hill here or there with tapioca cuttings stuck in it, or suddenly make the rows, instead of parallel, into a small box within a larger one, for no apparent earthly reason at all. But I must admit, it would break the boredom of some of their workdays! Suddenly 10 tapioca cuttings would be all in one place, and one 4-foot cutting laid down, instead of the normal 3- to 5-inch cuttings, which are laid in the row and covered over with soil. Actually I think it's all quite funny! I never know what I will see next! But they do seem to be able to work fast! I really admire their physical capacity for work. And it's so much fun seeing work done, by "Western" standards at any rate, in rather unorthodox ways.

Paniya men using "mummatis" (hoe/shovels) to prepare tapioca rows

Back to counting tapioca. As counting is more easily done when I'm by myself, I had a most difficult time getting rid of Kurian! Yesterday I didn't succeed, but today I explained to him that, while he does indeed help, it is my fault that I, nevertheless, do get mixed up unless I do it alone – especially since for uneven plantings I want to make small sketches of what is where. Sometimes I'm really embarrassed by his attentiveness. He didn't want to leave because he felt sorry for me if I worked alone. At any rate, aside from getting mixed up, which is quite true, I don't want a shadow.

Many of the little Paniya kids run around stark naked, while some others sling cloths about themselves and tie them in front of their chests.While most women keep wrapped up, some are completely bare-breasted, and others nurse their children quite openly while sitting, standing, or walking about. Often a small child is completely naked, except when carried by its mother, father, or siblings on a cool morning, enclosed within their wrapping.

Paniya group in front of 5-unit home

Groups of the Paniya women go into the underbrush near my hut and dig deep holes with their hands and with poles. I wondered why. I asked Kurian and Anthony to help interpret. They were digging different roots which they boil and eat. I was also shown some plants used medicinally for healing. I'll have to ask the Doctor what they are. Oh, I did try a small chunk of the tapioca tuber, peeled and raw, as so many eat it here. It was somewhat heavy in the stomach but no major problem and it does have a definite taste. I had to try what these people live off to such a large extent. After all, it must be edible since they eat it!

16 October 1966, Sunday

A letter was delivered to me while I was on the bus heading to a larger village.The postman, who now comes by here Mondays, Wednesdays, and Fridays, saw me on the bus at a stop. It was Thursday, so not his day to come. He got on the bus, gave me the letter and left. How kind of him!

Right now, more and more, I seem to be getting myself involved in all kinds of "social welfare" work (to put a tag on it). I give pills to nursing mothers to increase their milk. Men and women receive vitamin pills from me. The children receive wheat germ. Partly, I undertake "childcare". To mothers with babies with oozing eyes, I suggest washing the eyes gently with warm water. For children with colds, I suggest that they be kept covered and out of drafts. On Monday I had 10 patients! All treatments are at the suggestion of Dr. Narasimhan, on one of his visits down here from Kotagiri. The round trip itself takes him most of a day.

Last night I received 1000 acid-lime tree seedlings and 315 mangoes, provided gratis by the government. These need to be planted almost immediately. The monsoons are almost over, and thus it's planting time!

19 October 1966, Wednesday

Oh, just now outside, a high-pitched howling, long, and then interrupted by chopped-off barks. A whole pack of jackals! The second time I've heard them since I've been here.

Today, as I walked and talked with Kurian about encroachers, it came to my full realization that I am, in fact, the general overseer of the Paniya Colony now. Again, I'm so dense that I hadn't realized it nor thought about it that way before! Kurian mentioned that when he had been in charge, he had reported the encroachers to the police, and now that I was in charge, I should do the same. I think I'll "allow" the Doctor to handle this! Well, of course, I'm handling all the money now, approving or directing all the work to be done. But as I said, I just never put 2 and 2 together! Still, I'd be lost without Kurian, in spite of his somewhat dim-wittedness and our great language barrier. He knows where to get what, e.g. tapioca cuttings, and where to sell the lemon grass oil. Mainly we seem to get along fairly well. But concerning the things that need doing in this hut, he's hopeless! The washroom has no decent walls, just the sacks, which are rather transparent at night when my kerosene lamp is lit. No closet. No kitchen cupboard. And the "tea house" is still this hut. I prophecy it will require at least a few more weeks to finish the new tea hut, not the few days as I was initially lead to believe. As a result, I'm still with all my things in the little room, and the butler Anthony still inhabits the larger room. I'm not really worried, and it doesn't at all offend my ego. I'm sure that Anthony never before in his life has had living quarters more spacious than those of his "master". Oh yes, that's the only thing he will call me; it's his upbringing! So why not give an old man this little satisfaction, of having the larger room! Actually, when Doctor & Co. came last weekend, most of the group felt I should immediately move at least my sleeping quarters to a more stable structure, even if I left my things here and ate here. This was mainly in view of the storm which had come inside and the collection of insects they saw running and flying about my room, since there is no ceiling. And the "door" is simply boards placed in slots on the top and bottom, with some spaces between. But actually, it's not so bad! We definitely don't want to offend Kurian's feelings too much, in spite of the fact that he had rather mismanaged the finances, not altogether to his own disadvantage! Also, at least for some time to come, I desperately need Kurian's help. Furthermore, at least for a while, I've made up my mind to stay in this place because it somehow suits my fancy!! I must say, fortunately, I was brought some DDT to spray on my books. The other day in the morning, a colony of white ants (ravenous eaters) decided to use the floor of my room as their marching exercise ground! Some puffs of the DDT spray: in 15 minutes – no more white ants. On the whole, I let the insects be and rather use insect repellent on myself. So many insects are really just so interesting! And to think, I don't have to go out to collect them but can look at them right in my own bedroom. This would indeed be a rare treat for insect collectors! Right now several stray ants keep crawling over my bed, while I'm sitting on it. Nice juicy ones at that. Perhaps I'll have fried ants one of these days!

Re-thatching hut: Karimban (below); Urali (on roof)

My hut is being re-thatched (reinforced, actually) with more rice straw. Even so, the roof has leaked very little lately, as the straw had remained swollen because there has been some rain just about every day, both light and heavy showers. Also, the other day when I came home from work, I found that one of the Paniya women had cleaned the veranda, and it looked just so neat with its fresh coating of cow dung!

As yet I don't have a ration card, which is needed for rice, sugar, and wheat flour. The sugar I have was bought from some people who didn't use all their rations. Today I sent Anthony to Gudalur, with the once-a-day bus connection. It's a whole-day affair to take a letter for me to the Taluk supply officer, requesting a ration card. No luck, the gentleman was in Ooty. But I hope he will read my letter and act accordingly. These things may take quite some time. Actually I can manage fine. But it's annoying to have to spend Rs 3.50 for the bus trip in vain!

COSTS; FIELD HUMOUR; AND MY "QUACKERY"

21 October 1966, Friday

Although some fruits are expensive here, I can get plantains, as all the different sizes, shapes, and tastes of what we call "bananas" are called collectively. A small bright yellow-skinned variety, the length of a finger, but very thick is available at 7 for 12 Nay Paisa (nP = new paise; 100 nP per Rupee). Converted that would be about C1.5 cents. Quite cheap! And they taste very good. This area grows many types of plantain: ones with either yellow, orange, purplish, or green skin. Oranges are orange, yellow or green. The green-skinned oranges are plentiful and cheap right now. I have milk every second day, boiled only, as is all my drinking water. If one wants to convert, then my last week's food bill was between C$2 and C$3 ($CAN1 = Rs 7.5). A small loaf of bread costs 40 or 50 nP; 1 egg costs 25 nP. Eggs are expensive. Still, I eat a minimum of two eggs a day. I must not get that calcium deficiency again. One pound of beef is about 70 nP. I eat beef mainly on Mondays and Tuesdays, as Anthony can purchase some on Sundays, which is market day. While Anthony does cook mainly "western" food, once in a while, at my request, he does add more spices. A couple of times a week I now get curry and rice. Actually, I probably will be eating a semi-vegetarian diet, which is alright. Quite a variety of dishes can be cooked without meat. Fresh vegetables, especially cucumbers and tomatoes, are extremely cheap here! The cucumbers are huge orange beasts, but about the same taste as I am used to.

This week we started preparing a new tapioca field. I went out to the field in the afternoon to see how the work was progressing. I discovered that I had become the object of a practical joke. The rows had been made into the shape of a man with his arms akimbo, his hands on his hips; a stance I have been known for even as a child. Here in the tapioca rows, this was reflected in a 40- to 50-foot long "person". As soon as I noticed this, amidst all the more or less regular rows, everyone looked expectantly to see what I would do. Naturally, I had to walk from one end to the other before it dawned on me what it was. Then Vella, the self-appointed Paniya leader, smilingly took up the posture represented and burst out laughing. Here was indeed a field representation of the "new overseer", of me. It really was so funny and everyone laughed. Although I didn't ask, the artist was pointed out to me. It was Veran, a 55-year-old man. I think he was really anxious to see how I would react. As Kurian told me, Veran is known for hypnotism, and his family, at least in the past, for medicine and healing powers. I really never know what I will meet with next. It's so much more fun this way! Seriously speaking though, I hope my smiling and laughter doesn't encourage too many such pranks, for heavy rains could cause quite a bit of the soil to be washed away, since these rows run more up and down the hill rather than across in contours. I suddenly heard my grunts of partial comprehension of Kurian's explanations, in his rather poor English, echoed behind me. Two, not one, of the Paniyas were mimicking me! I turned around and when I saw them, I burst out laughing, and at once the whole Paniya work crew burst out laughing. I'll reiterate, I do admire and really enjoy the Paniyas' sense of humour!

Vella, self-styled Paniya Colony "chief"

I now have three patients under observation! Yesterday morning one man, Ganapathy, cut the side of his foot rather nastily with his "mumatti", a large wide type of hoe used also as a shovel. I saw him sitting there with his foot bleeding and quickly went to my hut to see if my first-aid kit had anything useful. I found an antiseptic ointment, some sterile gauze, and Band-Aids and was back to the "patient" one hour later. One hundred acres isn't small, especially if divided into several pieces and spread over the various hillsides. When I got back, Ganapathy was working again and had some plant leaves on his wound to stop the bleeding. Since I don't know the local natural medicine, nor what leaves were used nor their efficacy or potential side effects, I removed as many of these as I could and applied what I had brought. Although he wanted to continue working, I insisted he not work that day. I did pay him in the evening. After all, he was injured on the job. Later I washed his wound. So much clay and dust on the legs of the Paniyas! I also gave him an aspirin for the pain. Today we let him do light work. Yesterday afternoon, an older man, Punnadan, whose leg was badly bruised the day before on rocks, received similar treatment from me. So much dirt to wash off! Then tonight one more, Urali-Vellian, cut the sole of his foot. I gave him my sandals to wear, which were several sizes too large for him, much to the amusement of the others. However, due to the muddy roads, they served their purpose. He wore my sandals again today. So now I have three men walking about with white bandages somewhere on their legs. I hope the wounds don't get infected since there is so much dirt. I couldn't possibly get it all out of the wounds! And I wouldn't know what to do with infected wounds. Anyhow, I need to ask the Doctor next weekend. Since yesterday, I have given three nursing mothers special vitamins, twice daily after meals, to help them produce sufficient milk for their babies.

It's happened that two of the little Paniya children have cried when they've seen me. Now, some people would no doubt say that it was due to the fact that I'm a stranger; others, that they are frightened by my unshaven face – I do shave, only not every day. It somehow tickles my fancy, though, to say that they cry because my skin is white! And I'd challenge anyone to disprove it. Who can tell the thoughts of a toddler who can't yet speak? But now, as several mothers with their children approach me, one little one accepts quite peacefully her mother handing her to me to hold. In fact, the mother seems quite proud and happy that her child does not cry anymore when it sees me. She now hands her little one to me whenever an opportunity presents itself.

Phiroze Ghandy, accompanying the Doctor, brought loads of bananas for me to hand out to the children the following day. The Doctor and his associates all want me to win the affection of the children by little bribes of goodies! Next week the doctor will bring clothes for the three colony-born babies, who are all naked. One is just ten days old, born the day before I moved here. Dr. Narasimhan wants me to hand the clothes out. Back to the bananas. Loads of fun! And actually a good number of adults turned out for the "hand-out". It didn't at all feel like "charity", just making someone happy by giving something special. Actually, whenever I pass either of the two places where the Colony houses are, I invariably somehow or other joke with the kids. One boy once had an interesting type of flute with a buzzing, reedy sound (called a "Cheenam"). I asked him to play, and then I asked to try. I just tried to get some reasonable sound sequences, since at first it was hard to tell whether the next finger opened would be a higher or lower tone. At any rate, without intending it, I drew quite a laughing and giggling audience of children and their mothers. This was also a happy time.

Oh yes, I am introducing an innovation. I am asking each worker to place his thumbprint by himself in the square I point to in the official payment ledger book. Up to now, Kurian took their thumbs and pressed them in the appropriate spot. I started this new system yesterday. This was accompanied with considerable amusement. Some made much too feeble marks, others big blotches, and some went entirely into the neighbouring square. Some really needed to be assisted. If they did it by themselves, only the tips of their thumbs somehow managed to get on the paper. But today, some improvement! I feel that even in such small things one must help and encourage them to gain confidence in their own abilities.

SCHOOL BEGINS; AND SUCH LANGUAGE PROBLEMS

27 October 1966, Thursday

My newest endeavour – trying to get the children to school. During the Doctor's last visit, Kaima, one of the five oldest Paniya men, approached us and indicated both verbally and most demonstratively with signs, pretending to write on his left hand, using his right hand as pen, that he wanted the opportunity to send his young son to school to learn to read and write. I mentioned that I would check out what I could do to help. To my knowledge, none of the Paniyas in this Colony, nor many anywhere, could read, write, or do math.

Things here have changed drastically since last night! At the beginning of the week, when I first talked to the Headmaster of the nearby public school, which has 75 children but no Paniyas, he stated his willingness to cooperate, but he kept saying that the Paniya children wouldn't come! While school admission officially ended on September 23, the headmaster was willing to cooperate. But I had to promise to get the children washed and cleaned and to get proper clothing for them. The other children would laugh at them and scare them away if they appeared in their rags.

Wednesday evening the headmaster came to my place and we asked parents about their school-aged children, 6 to 11 years old, for elementary school, and drew up a list. So, 20 are on the list; their parents seem willing.

I bought soap and combs the same evening and sent the Doctor an SOS-letter up the hills, informing him about the clothing need and that the children would not be able to keep attending school in their rags/cloths. I provided approximate ages for each of the girls and boys.

Today I provided soap for cloths and bodies, as well as combs. Under the direction of my butler, Anthony, who really is helping me a lot and seems to find great enjoyment in this endeavour, and one of the Paniya women, the prospective school children went to the well and washed their cloths (or rags). Other mothers only watched from their houses! I requested several of the older "school children" to draw water in a bucket from the well. Then we soaped them one by one and dashed water all over them. Most let this be done, willingly. It has been a hot day. Mind you, I had to catch a few first! For some this was probably their first soap bath in their lives! Having finished with the school children, I had all the little toddlers bathed too. Only one of the school-aged children won't be going to school. She has to look after her little brother, as they have no mother.

Still, tomorrow is to be the first day of school and I hope in the early morning to at least get them partly cleaned. But I am afraid that their one cloth wrapped about them may easily cause trouble at school. So things are hectic these days!

30 October 1966, Sunday

On Friday morning, the first day of school, and with help from Anthony, I washed dirty faces and oiled and combed the school children's hair personally. Then I showed each child the mirror. They had the most somber expressions when they looked at themselves. It may have been the first time they had seen themselves in a mirror. I then called each child by name and issued a slate and slate pencils, as my present to them for the start of school. While I know most of their

Paniya children, oiled, combed and ready for school

names by now, I must make sure I soon know and can pronounce them all properly! Also, with the described transformation, some of the children were not that easy to recognize. Accompanied by Anthony, I marched my motley crowd of the cutest "rags" (i.e. children in rags) one can imagine, to school. I was so touched by the sight of them walking along so quietly, probably proudly, in anticipation. I was so worried about how the other children would receive them! After all, the Paniya children just had rags on, not proper clothes like the other children. This worry isn't over yet by a long shot! But one major hurdle is over. On our way to school, a local non-Paniya boy, who had befriended me a little while ago and who speaks a fair bit of English, loped along right next to me, as he has done several times before. I requested that he and the other school children be kind to these "new" school children. He agreed to this. It turns out that he is something like a class "senior" in one of the upper classes in this school. At school, all the teachers welcomed us most heartily.

At the morning assembly outdoors, the headmaster called out the name of each of my 19 Paniyas, the first and only Paniyas in his school. Some of these little Paniyas stood quite awestruck and didn't budge when the headmaster called them up, until I nudged them, to the amusement of the other Paniya kids. They seem to be amused at almost anything I do to or with them! The teaching staff was rather bewildered and marveled that I actually knew the names of most of these children. The thought struck me that perhaps not many people in this region know Paniyas personally! I must learn their language! Tuesday evening my Malayalam lessons, given by the headmaster, will start. At noon, the children were barely home 10 or 15 minutes to eat a small lunch, before they rushed back to school. When some of the fathers told me that the children had gone back to school, I checked in the school yard. No familiar faces there! The headmaster asked some of the other children where they were. Indeed, that is how we found out that all the Paniya children were in their schoolroom, sitting and waiting quietly for class to start again. The headmaster, the other "masters" (teachers), and I must obviously have shown our surprised bemusement, for soon we all, including our new Paniya students, were laughing heartily. The fathers seem to me to be watching with great interest and anticipation.

I had to agree to take full responsibility for the Colony Paniya children. Thus I applied for their school admission as their official guardian. Actually, it's a terrific responsibility and I hope I can live up to it. I really want to. I think this educational aspect of the Paniya Colony project may be as important as anything else I could possibly do to help!

That evening, Anthony lined the school children up in front of my hut and at my request, explained to them a German custom. When I started school, after the first day, we all received sweets and perhaps some other little goodies! So now, calling each by name, I gave him/her a handful of candies. As it was also time to pay the men their wages, all the fathers were present, as well as quite a number of the mothers. I could read in the parents' faces quite an interest in what I explained (through Anthony's interpreting) and what I did. Well – the first day of school had gone well! What will the future bring? It is encouraging to know that the inspector of schools, who visited the school the day before the Paniya children attended, was very much interested in this "project". He is recommending that clothes be issued to them free of charge (since the Paniyas are very poor indeed!). I also wrote the Doctor to see what he could do.

I am really so thankful to God that up till now it has gone so well. Of course, this whole week was somewhat exhausting! With the way things here began to snowball these past two weeks, I feel like one of those "fools" of "Fools rush in where angels fear to tread". Probably every morning after the fathers go to work, I will ask the kids to come for cleanliness inspection and hair combing, and then accompany them to school for some time. If not all, I sincerely hope most will continue on! But even if only a handful remains, this will be quite important to this Colony, and make me ever so happy.

The language problem is really my very biggest concern here. Kurian, as my interpreter, does not understand enough English for me to explain things to him in detail. He finds it quite adequate if he understands the reason for doing something which I suggest. He is happy he got it and that the Paniyas then do the correct thing. He has, however, no intention of passing the reason on to the Paniyas. Also, if I want to try to admonish the Paniyas, or a particular Paniyan ever so slightly, Kurian scolds and threatens them with reduction in salary. I tried to point out to Kurian that this is not much different from the former landowners of the Paniyas. Somehow he does not "get it" that there could be a different way. I don't think this relates to his lack of understanding of English. And that is precisely where I miss not knowing Malayalam the most! I want to be able to explain to the Paniyas directly what my concerns are, in my own way. Not that I know the "correct" method of dealing with the Paniyas, but I would like to have the opportunity to see their reaction to my explanation. Then too, many of the Paniyas do want to tell me things, explain things to me, ask me things, and indeed they try, but alas, I need to repeat that I cannot understand them ("menesillai illa").

Often questions from the Paniyas via Kurian do not make it to me, as Kurian decides whether they are important enough to pass on. A good example of this took place a few days ago. Punnadan, my second patient, stood in front of my room for quite a while. Then he talked to Kurian. Finally I asked Kurian what he was after. Kurian laughed, telling me that the Paniyan asked for Re 1 to buy soap for his three-week-old baby, to give it a bath. Kurian thought this was completely useless. Well, in my "First Aid Kit" I found a special disinfecting soap, which I then gave to Punnadan. Having used the soap for the purpose intended, he came back today to return it to me! On the Doctor's most recent visit, Vella suggested to him that Anthony should just translate for me directly with the Paniyas. Thus Kurian would not be required. I think this suggestion reflects more on relations with Kurian than on anything positive I may have said or done.

Anthony, a non-Paniya, cooks only for me. But actually he cooks for me as though for two for I eat the same meal at lunch as I do in the evening. He cooks for himself separately. Kurian, another non-Paniya, sleeps in one of the rooms in the Paniya houses; thus he has a roof and four walls.

Everything is cluttered in my little room in this mud hut. I constantly have to move some things to get at something else! There is no space to walk about. If enough is removed and shifted, I can mainly just sit in the room and have everything within reach. Also, I have medicines everywhere! This past week I really had the opportunity to use most of them for my patients, at least several every day! Ten on one day and six on another.

PERMANENT BUILDING PLANS; OFFICIALS VISIT; AND A CELEBRATION

Without doubt, I need to reflect on what I am doing here. I do sincerely hope that this program works! But what if it doesn't work, and I am somehow contributing to its failure? Of course, I will still have had a terrific experience!

NAWA is planning a permanent set of buildings here. The first building would include living quarters and a dispensary. I really see the need for a nurse here. This quackery of mine for the past two weeks can't go on forever! Also there are many types of sicknesses and ailments I really haven't a clue how to treat. Nor the time! The building hasn't been started yet, but a contractor has been hired and will start soon. But I am really happy that I will have had these few months at the beginning to be the only "European" (i.e. Caucasian person) working and playing (a lot – with the children!) with these Paniyas. And already now I feel strongly that at some time in the future I must really try to take the time to re-visit India and particularly this Paniya Colony!

11:30 p.m.: About 12 hours ago I was interrupted in my writing when Doctor and Miss Armstrong arrived. It was an exceedingly busy day. (Just now there's one of those jackals yapping directly outside my hut!) Miss Armstrong brought clothing for almost every one of our school children! You can imagine the cheerful faces! Her sewing machine was going through most of a night after my letter to the Doctor had arrived in Kotagiri. She had sewn a blouse and skirt for each girl, and a shirt and short pants for each boy. Tomorrow, the second day of school, all Paniya children can attend in clothes acceptable to the headmaster!

The District Welfare Officer, responsible for all of the Nilgiris, paid a visit, along with the District Agricultural Officer, the Agricultural Extension Officer, the Revenue Inspector for Tribal Welfare, all their

Paniya children ready for school: with new clothes and slates

assistants, and Kesavan, the 23-year-old local silk worm/mulberry farm manager. The Doctor embarrassed me no end when, in showing them around and explaining how we were trying to help the Paniyas, mentioned: "Mundel (this is what he calls me) has been here just a few weeks and already one can see progress in the organization of the work." I was particularly embarrassed because I had made several faux pas! For example, I had planted the mangoes 20 feet apart rather than 40 feet. I had followed the suggestion of the local representative of the Agricultural Demonstration Ministry! Well, apparently he had misinformed me! Anyhow, now we will have to dig up every second tree and replant it! What a nuisance! Also, since there has been no rain for a week, we'll have to water all planted trees individually with water from the well. This means running up and down the hillsides for all the 1,375 trees!

This evening the Paniyas were having a big celebration. I joined them! They were playing flutes and drums (called "Thudi" by the Paniyas), beating the drum with the curved part of a stick, while children and women danced, a rhythmic, slow-motion "jerk-like" dance. It was all very interesting. In the moonlight, I sat with a flock of Paniya children crowding around me. One baby was placed on my lap by its dancing mother, who had danced with it on her back, wrapped inside her cloth. It was quite thrilling actually, these Paniyas, dancing and making their own music. I'll see if some day they'll dance and play in the daylight, for a photo. Oh yes, I took my recorder along, my German Hohner "Blockflöte", and played along as best I could. The flutist and I also switched instruments, and I let several others try.

Paniya musicians

PO MONEY ORDER WITHOUT ID; "FOUND" BY "EUROPEANS"; MEASURING THE PANIYAS

10 November 1966, Thursday

Sometimes I get upset when I'm in the fields and find so many things done differently from what I had instructed, largely due to Kurian's interpretation, his "better judgment", or, I feel, his obstinacy.

I have donated Rs 350 (C$50) to NAWA to help meet some of the present financial shortages. This money is out of my yearly CUSO allowance for "extras", such as travel, gifts, etc. Also, as the partial "parent substitute" for our 21 school children, I think I'll accrue more expenses as time goes by, for things such as slates and slate pencils which break readily. I have heard so much about difficulties with postal money orders (M.O.) in this country, both sent from abroad and within the country. Well, the M.O. for my donation came from the CUSO-Delhi office. I had to walk to the post office, a nine-mile round trip, not six, as I had previously thought. This takes quite a chunk out of the day. Once there, my first time, the postmaster greeted me and asked me to sign a paper and counted out the money. He didn't even ask who I was. I had never met him. He didn't ask for any form of identification! Actually, I'm the only "European" in this postal zone! Since the money orders work, I will not try to open a bank account. The nearest place to open an account is in Ooty or Kotagiri, with a one-day round trip by bus just to access it! Not very practical for me.

This brings me back to the "Europeans" of the tea estate. Since Mike Congreave, the 22-year-old Assistant Manager of a nearby tea estate, met me on a country road, I have been considered "found" by them. They said that I must have been quite busy to have been able to have been here one month without any of them seeing or hearing of me. Actually it shows me how little communication there is between "Europeans" and Indians. Numerous Indians knew very well who I was and where I was from, how long I was here for, and what I was doing etc., because they had read about me in the Madras Mail, a

Tamil-language newspaper in this State. And this doesn't include the local "jungle telegraph" which seems to spread news everywhere! As for being "found" by the "Europeans", somehow I can't help feeling as though this is a real encroachment on my privacy! I had no "need", just yet, for being "found" by "Europeans"! But Mike seems a nice "chappy".

Monday is "Children's Day" (Nehru's birthday) at school. I will make a small contribution on behalf of our Paniya children to purchase sweets for the "function" at school. After all, their parents can't afford it! Naturally I'm invited! These school children need so much time, care and attention which they aren't getting! And I can't possibly do an even partway adequate job in this connection! And I want to so much! Naturally, I get to sew on buttons and sew up torn shirts. But I must say, Anthony really helps me too! Then the clothing needs washing, etc. etc. Several times now on the way to school, the children swarm about me and several take hold of my fingers – this way more can be accommodated than were I simply to grasp a hand. Some catch hold of my clothing. Of course jealousies arise. These kids are so cute and so friendly and open. But, it's not easy for a 24-year-young bachelor to look after 21 youngsters, especially since one could easily devote all one's time to it! And the fields and men also require full-time attention. I fear as a consequence both may suffer!

Oh yes, I'm trying to weigh and measure the height of all the Paniyas in our Colony. I've finished with the school children and the men. They all enjoyed it. Some whom I didn't get to on one day were already in line the next, one standing on the scales, by the time I was ready. I was actually much amazed at the small average height of the men! They somehow seemed taller to me. Their average height is 5 ft 1.2 inches! So, even this group of Paniyas verifies the literature which states that the average height is between 5 ft. and 5 ft. 2 inches.

As Mike C. keeps rain records, I now know that here on the western slopes of the Nilgiri Hills we receive about 140 to 150 inches of rain per year. More than I have ever imagined experiencing! One evening he invited me to dinner in his bungalow, right on the edge of a hill, all under tea cultivation. He lives alone and has three servants. At his house I saw electric lights for the first time since I came to this Colony! He has a generator just for his bungalow.

This week I started Malayalam lessons, provided gratis by the Headmaster of the school our Paniya children go to. Actually, I'll also be taking notes from the children. They seem to love to teach me!

The most common greetings I get here are, "Good morning", regardless of the time of day, and "Salam". There are Muslims in the area, but now all non-Muslims also use this greeting.

I must go to Kesavan's place because tomorrow we want to leave early at 6:00 a.m. to catch the bus to Calicut (Kozhikode) in Kerala, where we will attend, with the Doctor, the Diwali Festival days (a Hindu festival of lights). I really didn't know how I could manage the time away from this place, but this also being a school holiday, one series of problems is solved. Then also it's raining these days so another series of problems is solved. I feel I can safely leave Kurian in charge.

CHILDREN'S DAY; WAGES, DEBTS, AND RESPONSIBILITIES

17 November 1966, Thursday

As I sit on my bed, the smell of cow dung is filtering into my room. My veranda is again getting a fresh cow-dung surfacing!

On Monday, November 14th, Nehru's birthday and Children's Day at school, I contributed Rs 5 for sweets for the children. There were no classes that day, just a brief celebration, which I was asked to attend. Assembly took place around the flagpole and apparently because the Paniya children had requested it, I was cordially requested to hoist the flag of India. For the occasion, the flag was filled with paper flowers, which scattered all over me as I pulled the cord, once the flag had reached the top of the pole. The children were all lined up to the left and right of the pole. Then followed some response songs by the teachers, with an occasional child leading and the rest responding. A short speech on Nehru by the headmaster followed in Malayalam. I caught the name, no doubt due to progress in my language studies! Ha! Ha! Then sweets were distributed. I was asked to start this part

School assembly

of the festivities, which for many, if not most of the children, was no doubt the main part!

Reflecting a bit, perhaps I do in fact have some hermit tendencies! I feel this loneliness more when with "Europeans", as for example, last weekend in Calicut. There I unexpectedly met six CUSOs, one Peace Corps Volunteer, plus our host, an Indian and my travelling companion, Kesavan. We briefly also met five other Peace Corps volunteers. With only about a half-hour, it was too short a time for anyone to get at all acquainted. Overly cheerful young "Europeans" somehow depress me, at least at this time. I may feel different in the future. And I don't at all feel "at home" with these plantation managers near here, who now have "found" me. I spent an evening with them at their club. But I don't have to see these people too often. I'm sure that most of the Europeans in the area would never understand how I could possibly prefer my isolation to their company. My "hermitage", however, is not at all devoid of people! As a matter of fact, it's literally swarming with them at times – mornings, noontimes, and evenings. The tea shop is still in the back of the hut! In the evening when wages are paid, many of the Paniyas are here. In the mornings, the children visit before school. Of course, these people are of a different skin colour and speak a different language.

There is so much I don't know yet about the land! Just the other day, Kurian showed me some 10 cashew-nut trees on this Colony farm I hadn't noticed before. But then, I didn't know what cashew nut trees looked like either! As for silk worms, we won't be rearing any for awhile, since the mulberries, which were just planted, need plenty of leafy growth so we can pick the leaves and feed the worms. Perhaps during the latter part of my stay we'll begin with the actual silk-worm rearing. This will depend on the growth of the mulberries.

Some matters which would seem trivial in Canada or the US can be rather major issues here, for example, the spending of a few rupees here or there, when the Paniyas receive only two rupees a day – the amount probably needed to provide daily food for their families. So the emphasis on money here centres on overcoming starvation and not on expressing wealth through the purchase of things. But still it annoys me to have to be so much involved with money here, such as the paying of wages daily. I hope to reduce the frequency of paying, if it will work! After all, the Paniyas must learn to plan ahead somewhat before they can take over the running of this place! Silly money deals abound, which for the sake of the welfare of the Paniyas I have had to make it my business to try to settle! For example, numerous Paniyas still owe former landowners between 5 and 25 Rupees. Some of the Malayali-Chetti landowners came to ask me to collect this by deducting it from their wages. I asked the Paniyas concerned how much a week they felt they could pay. I also asked them which day would be most convenient for them, and not the landowner. Then I asked the landowners to come to my hut at paying time on the agreed-upon day of the week, assuring them that I would see that they got their money. Kurian first asked a Chetti when and how much should be repaid, but I insisted that the Paniyas be asked, even though Kurian laughed at these "antics". The Chettis must also have thought it strange! However, I feel that even such small things as letting the former semi-slave-labourer Paniya decide which day is most suitable for him, and to do this in the presence of his former landowner, whom he rather fears, is important in helping the Paniyan gain self-confidence and a sense of human dignity! Of course, I don't really know if in the end it is the demonstrated independence of the Paniyas or the "power" of the white skin, which made this small step possible! I feel in all my dealings that an understanding of and training in psychology would be ever so helpful! I act rather on instinct in my dealings with people, as an animal would, and on some past experience, both actual and learned through readings and discussions, but without necessarily any real deep insights.

Let me reflect a bit here on responsibilities, and my situation, where the language problem is so severe! These Paniyas, who two to five months ago were still living as bonded labourers, quasi-slaves, are expected to shortly be able to run this Colony in a cooperative manner! And at present I still pay them daily wages! But even such seemingly small steps as changing this to paying only every second day, and gradually even less frequently, are important, I feel. After all, learning to plan a few days ahead is essential if eventually one has to await a harvest for income! A long ways away, I think! Dr. Narasimhan hopes to achieve this in about one and a half years. Having been here such a short time, I have difficulty seeing how this can be. However, the Doctor has known the various tribal peoples since his childhood, over 50 years ago.

As my lamp is running out of oil, which is stored in my butler's room, and I don't want to wake him, I will close for today.

BARBER; OFFICIALDOM PROMISES?; EARRINGS

20 November 1966, Sunday

Some more about the school children. Today I took all of the boys who needed haircuts to the barber. There was much questioning and looking by all as we passed! The barber was quite flabbergasted and turned away several subsequent customers, as he now had his hands full with eight boys all at once. He gulped down his tea and set to work, following my sign instructions quite well. I don't think any of these boys had ever in their lives been to a barber before and several had never had any of their hair cut. All seemed quite keen. As a matter of fact, two of the Paniya men also showed up at the barbers, having apparently heard that the "Master" was calling people for haircuts! But I was hard as I feared having 28 to 29 men requesting me to pay for their haircuts, which cost twice as much as those for boys! Afterwards I invited all the boys for tea at the teashop just across the street from the barber. Although they said a sentence requesting money for tea, they didn't expect me to take them up on it. I think they rather said it because they knew that it was one of the two things I could understand. Those few things the Paniya children know that I understand they repeat and repeat with great mirth, for example "good morning" (or rather: "goojde morrning"), "salam", and "twenty-five". I learned from them in turn to say "irupadh anju" for 25. One of the boys especially, Vellikan C., Chunda's son, kept repeating and I did too: "irupadh anju, twenty-five; irupadh anju, twenty-five". So, all the boys sat on two benches very cheerily munching their buns and drinking their tea. Here I did include any misinformed Paniya men who happened to be here. Then, home we marched, quite a joyful group, and really looking quite smart, with cleanly cut and oiled hair! I'm sure that many of the local people are convinced that this 'saibu'[6] is rather unpredictable. Actually, they're right! The other

[6] Originally, it seems, used in Tamil for Muslims, but here as a corruption of "sahib", used for "Europeans"

day in Cherambadi several boys turned to me and said straight out "Colony Saibu". So it seems I can be identified without people knowing my name! When I was still at Kotagiri, prior to moving to the Paniya Colony, my attempts at being called Henning were in vain. Well, I guess I'm an ethnocentric clod, because such things seem to be unheard of under these conditions. Really, I'm not being a snob by letting Anthony and Kurian call me "Master" and "Sir"; and the Paniyas, as well as many in the local population refer to me as "Saibu". What really matters is how my interpersonal relationships with these different peoples are. And I try where at all possible to take each as an equal. Mind you, I'm biased for the Paniyas, and subsequently am occasionally too hard on Kurian, which I do try not to be. I am definitely hard on former landowners whom the Paniyas worked for, and on encroachers. In a way I feel I have to become a lion, to growl at these people to leave the Paniyas alone and to accept them as full human beings. At least these are my feelings. How well my actions manifest these feelings, I can't assess myself. As it turns out, the only ones calling me Henning are the Doctor & Co., Kesavan, and Mike. In our games of words and phrases in Malayalam, I asked a group of the Paniya boys, "Ente peru endh-anu?" ("What is my name?"), to which they replied they didn't know. So, jokingly I suggested "Paniya Colony Saibu". It must have been quite a joke since it was repeated till all had grasped it and burst out in laughter. Their laughter seemed to indicate that they agreed that this was a fitting name.

In my work here there are so many problems – in the daily work and long-term! While these indeed present a challenge, they can also be frustrating! In the last few weeks, I think I've been suffering from symptoms of so-called "culture shock". In other words, things gone wrong have annoyed me more than perhaps they should have. Of course, I do have so much responsibility for the goings-on in this Colony that I can't help but count as personal failures work done poorly or incorrectly. After all, some of the errors done now will bear consequences for years to come! I also get annoyed at numerous officials, who, although expressing "great interest" in this Colony, just talk and don't seem to do anything! While I do understand the proverbial "This is India!", does that mean I have to watch while things don't get done or delivered here on time? We still don't have the fertilizer which I've tried to get since I arrived here. Now, for some days there has been no rain. Perhaps a couple more rains, and then, bingo! The dry season will be upon us, and it will be too late for the water-soluble fertilizers. The Panchayat Union hasn't supplied the additional clothing for the school children! The Agricultural Extension Officer (AEO) still has not sent his field man to take soil samples, which were to have been taken the week I got here! Subsequently I took some soil samples myself and let the Doctor take them to a lab. Neither has the AEO sent the engineer to mark off the slopes for major catch ditches for water ("bunds"). The Special Revenue Inspector has not sent the veterinarian for weeks now, to certify a milk cow to be donated by the Harijan-Welfare Department to one of our Paniya men! Because of the many different things I have to look after here and my extreme immobility, I'm rather tied down! Last week, because of urgent work in Kotagiri, there was no Doctor visit, and hence no money for the Paniyas' wages. A little came by money order, but not enough. I had to quickly decide to go and see where we could sell our lemon grass oil. This gave me enough for wages for about two days. By then, almost all my own money was also used up. Of course, I was refunded what I had paid out when Miss Armstrong and Miss Martin came with the money. I really am being tested, living from hand-to-mouth. This exercise should create nerves of steel! Under such circumstances, it's futile trying to plan days ahead!

I think I'm learning my lesson about making appointments with various people to come here, who then subsequently don't show up. So now I try to go about my daily work as though no one were to come. If they come, anyone can tell them where to find me, or else they can wait. I also visit the various officials and others in the 'Taluk' headquarters of Gudalur by bus, just hoping some will be there. Most, of course, have letters from me from two weeks back, but have made no response yet! Since I didn't come here to play a game with officialdom, I must go to their offices. However, I never know what to expect back here when I return after even a few hours away, let alone a day. Tomorrow an engineer may come to inspect the foundation of the new house, which is not yet completed. Well, if he comes tomorrow, he won't find me here! On the other hand, I may meet him on the road coming in a Government of Madras jeep. If that's the case, I could have the bus driver stop and let me off.

The scenery here is absolutely beautiful, with mountains at varying distances in all directions. In three directions are the mountains of Kerala and in one those of Madras. It's so beautiful in the evening to see trees silhouetted on the slopes of hills against the deep orange skies of the setting sun. The country is hilly, and the roads very curvy. This point was really brought to mind when one day Anthony and I were walking and he commented on how long a certain straight stretch was! Actually it was less than a quarter of a mile. But I guess there are not many such long straight stretches around here.

Aside from the women and girls, there are even many young boys who have small pointed wooden pegs through their earlobes. A type of earring many of the Paniya women wear consists of a wide roll with the red part of black and red seeds[7] exposed on a beeswax background, all about 1.5 to 2 inches in diameter. This size is not exaggerated. Actually it looks quite nice, red seeds on a dark background, surrounded by light-coloured dried palm leaves with a deep brown margin. But what is so interesting, and which I only found out

Paniya women's red berry-filled earrings (actual size)

[7] Red seeds from the legume, *Abrus precatorius*, called various names such as Precatory bean or Rosary pea. The top two-thirds of the seed is red, the lower one-third, around the hilum, is black.

about two weeks ago, when one woman showed me the earring, is that the earring doesn't hang on the ear, but is in it! In other words, the deep brown margin is the lower part of the vastly enlarged stretched earlobe. I could hardly believe my eyes! But it's true!! Of course, since most of the women have their ears completely covered by their mops of hair, one could at best see the red seeds shining through. The pierced ears of the Paniyas are even annotated in the 1961 Census of India: "The women start dilating their ear lobes from the twelfth to thirteenth year. They first wear small scrolls and they slowly increase the diameter of the scrolls, till they reach the breaking point." Yes, I've seen several broken ones.

All the High School Students from the area around Cherambadi, the larger village but without a high school, must walk the six miles each day to go to school in Erumad. That is twelve miles, counting both ways. Good exercise, eh?

In local "hotels", not places one stays at, but restaurants, banana leaves serve as plates. What a most useful device. No dish washing! Just throw it to the dogs, literally, as dogs wait for all the food scraps. Several times now, Kesavan and I have had some evening snacks at the corner where we part. Each of us has just over a mile to his place. There we sit in the tea shop drinking coffee or tea and eating some biscuits (cookies), with one candle placed before us. We are usually the only guests at closing time.

Off to Gudalur tomorrow. However, there may not be a bus. It very often breaks down!

21 November 1966, Monday

Kayyunni corner, waiting for the bus to Gudalur

This morning I asked the children to walk to school by themselves, as I wanted to organize my things in preparation for my trip to Gudalur. On passing one of the Paniya houses, I saw one of our little school girls, about six or seven years old. She was naked, as is usual for the little children when it is warm and they are not in school. Her smaller brother was wearing her dress, and I asked her kindly to come along. I told her I would walk her to school, which was on my way. Without saying a word, she pulled her dress off her brother and put it on herself, went to her home and locked up, came to me and took my hand, and held it tightly until we arrived at school. I was rather touched. I assume the little brother went to one of the other Paniya homes. How I would like to be able to do so much more for these little children! But I don't even have time to sew on all their torn buttons!

RESIDENTIAL SCHOOL CONSIDERATIONS; ENCROACHERS

27 November 1966, Sunday

Paniya Colony

Yesterday afternoon the District Welfare Officer (DWO) came to the Paniya Colony with his entourage. I had a number of items to discuss with them. While I feel that it is good for our Paniya school children to spend time in school with non-tribal children, I initiated a discussion about possible arrangements for the next school year, to enroll the Paniya children in a tribal residential school. There they would receive gratis, full board, all needed clothing, and full school supplies. Also their attendance would not depend on whether or not a parent were working, i.e. whether they needed to stay home and look after smaller siblings. And if the mother wanted to visit relatives in a neighbouring village, she would not have to take all her children along!

Later the Doctor and his entourage arrived, and I had much to discuss with him, not having seen him for three weeks. This week, once again, I had considerable problems with the encroachers. I have started our men on digging pits for future trees mainly to hinder further encroachments. Many of the people who had been using this area in the past for grazing were quite vociferous in their protests. One who said he paid tax on the land produced the papers on my request. They were in Tamil, of course. I am being considered unreasonable for not allowing them to graze their "few" cattle. But, so many have only a "few"! Now it's dry season, thus soon there will not be much grazing for our own cattle! The land is largely overgrazed. I can't allow these encroachers to graze their cattle too! Our Paniya Colony already has 10 oxen and two water buffaloes. Through the "jungle telegraph" I learned a "grazing party" was planned, to see what I would do. I was met by quite a group of men, led by an English-speaking troubleshooter, apparently the President of the Cultivators in the area. I explained the situation. I fear in the long run this one man will cause me some more trouble. One of his arguments was, "Whoever heard of a tribal scheme that has so far succeeded!" Of course, he may be right! But that is no justification whatsoever for the encroachers continuing to use Colony land. While outwardly it is easy to be hard, it really bothers me, since my skin is white, and thus my actions may be misunderstood! I certainly can understand their plight too! They really don't have enough land for grazing their cattle. When all these men came, with their troubleshooting President, I somehow felt as though I were meeting emissaries from another country!

OFFICIALS MET; MARRIAGE BY ELOPEMENT; NIGHTTIME MUSIC

29 November 1966, Tuesday

Earlier, I mentioned several officials whom I had hoped to meet in Gudalur. While I met all but one of them there, during the past week all have been here at the Colony to visit me! Coincidence? The BDO (Bock Development Officer, who is also the Commissioner of the Panchayat Union) came with the engineer who was to inspect the foundation of the new house. The BDO told me that what the encroachers had shown me as payment of tax for use of the land was in fact a fine by the government. He also talked to some of the Paniya men, apparently convincing Vella that Tamil would be of more use to the children than the Malayalam they are studying in school now. Tamil is the official government language in this State. However, as this area borders Kerala, the population here mainly speaks Malayalam. Well, let's see what happens in the coming weeks when the Tamil teacher arrives! The BDO did not yet bring any of the clothing for the children! Who knows when they will arrive!

While still in Canada, I had read about the not uncommon "marriage by elopement" among Paniyas. Well, here something like that happened, but not with unmarried people. One of our men, Karimban, left his wife and took another one from our Colony. The two spent two to three days somewhere in the jungle, with bow and arrow, and then returned. Before they entered the Colony grounds they came to my hut. After standing there for some time, Karimban approached my door and requested vitamin tablets. As he had been gone for a few days, he had not received the multivitamins I distributed to the working men daily! Well, now he lives with his new wife in the Colony while his first wife moved in with his sister, Chunda. Since his first wife now has no one to provide for her, I try to find some jobs for her daily, so that she can earn some cash. Karimban's new wife had previously been married, but her husband had left her, so she had been living with her parents and siblings. Karimban is a brother of the "chief" Vella. Another brother is Urali. A week ago Urali and his wife had a quarrel, during which he placed her hand directly into the fire! As interesting as I thought it might be to be a doctor here, I'd never make a good one! With all the suffering I treat, I almost feel as though I were the patient myself, although I was never very squeamish. Here I have learned to be even less so! I had to interrupt my writing as a mother with a scabies-infected child came for treatment by "Quack Mündel"! Now the child has been treated, and after an injection by the Doctor on Saturday, the scabies is diminishing quickly with my twice daily application of strepto-penicillin ointment. I hadn't seen the Doctor for three weeks, so he greeted me heartily, taking my hand in both of his.

About once a week I visit Kesavan and observe how the silk worms increase in size from moult to moult. Four times daily, fresh, finely chopped mulberry leaves are sprinkled in a thin layer over them.

Two and a half hours after my visit to Kesavan, I am sitting at the corner tea shop, where Kesavan and I have several times taken snacks in the evening, the closest tea shop to my hut. The tea shop in back of my hut is finally closed because of a quarrel between Kurian and his tea maker. I am waiting for the District Education Officer (DEO) to visit the school today. He is coming down from Ootacamund (Ooty). I want to find out what is happening to the sanctioning of midday meals for our school children.

One evening last week, I had to see Kesavan concerning compost and manure. Since he wasn't home, I had to walk all the way to Erumad where he was having dinner. An evening stroll of eight miles! On the way back I heard some Naika (one of the hill tribes) singing and shaking their rattling rhythm instruments. Walking further on in the moonlight, I could now hear singing from the direction of our still distant Colony. This was the first time I had heard regular songs from the Paniyas. Soon I reached the road just below the group of two Colony houses (the other group has three houses with five family units, each consisting of two tiny rooms). I stood and listened. Suddenly I was spied. The singing stopped! Some of the men came closer to see who it was. Then "Durra! Durra!" (Master! Master!), and they waited for me to come over. Vella asked some of the others to move aside and brought a mat for me to sit on. Then the singing and dancing by the men (mainly one or two), and quite often one of the boys, continued till 1:00 a.m. I was really quite thrilled!

Then on Saturday evening, the Paniyas put on a drama! Actually a musical comedy with much dancing. To the accompaniment of drums and cymbals, a lead singer sang and the chorus would repeat the lines. It was very funny, somewhat slapstick. Only the men were dancing. Indeed, they also danced the parts of women, having heavily powdered and painted faces. Numerous costumes consisted of quite a few of the girls' skirts and dresses we had provided for the school children! A leafy canopy, supported by posts, served as housing for the dancers and musicians. A bench was placed on the side. The Doctor & Co. stayed for the first half-hour; but from then on till 2:00 a.m., I was the

Bojan and family in front of their home

only one (as the "Durra" of our Colony) occupying this seat of honour. Vella even presented me with two packets of cigarettes. I was quite taken aback, and, even though a non-smoker, I felt I had to accept them! Soon I was surrounded by more than half a dozen of our school children, two climbing onto my lap. About half of the actors, dancers, and musicians were from our Colony, where all this took place, and half from neighbouring hamlets. I had given four men leave in the afternoon to construct the canopy and clear the ground. The audience consisted of hundreds of Paniyas as well as numerous other local people. Four temporary tea shops sprang up in the surrounding area and the audience spread in three directions from the "stage", everyone sitting on the grass. It was a real folk festival!! This was on the evening before a full moon. I hope to find out eventually what the action was all about. While it centered on a man, and a girl who didn't want him, at least at first, as the actions made clear, much had to do with agriculture. There was pounding and husking rice; then carrying money home; all in rhythm to the music. A collection was also taken up. Some individuals passed coins to the actors during the performance, and after the main act, a plate was put on a table. Individuals passed coins to a "caller", who very eloquently announced the donor, and the recipient – which actor or which musician was to receive the funds. Then the amount was mentioned and the money placed on the plate. While I was prepared to pay, I didn't know exactly when and how. At one point in the drama came a scene where a coin was needed. I saw this and asked Anthony. He asked one of the Paniyas, and during this interval the actors stalled for time so that I could understand exactly what I was to give the coin for! Well, just to be safe, I gave several coins, which were received by gleeful bounds by the actor/dancer playing the role of the hero. He immediately gave one to the "heroine", and the play continued! At the collection, I gave Re 1 for the main singer and each of the main actors, plus Re 1 in general. After the main act, I invited all of our Paniya girls who happened to be on hand for tea and buns. The boys had had their treat the week before after their haircuts. Here too the odd Paniya man wanted me to buy tea for him, which I did. Well, at 2:00 a.m., I left the scene. As I found out the next day, it continued for several more hours. Sunday was a day off! Good thing!

Earlier I mentioned the Naikas. It seems five years ago a well had been built for them but had never been used because they hadn't been supplied with a bucket or a rope. And they really don't have money to spare! So they had been using dirty water from the rice paddy fields. Through an interpreter, I asked them if they were interested in a new bucket and rope. I asked them to fix a beam across the well and attach the pulley. The original beam had been eaten by white ants (termites). Within about two days everything was ready. I sent Anthony to buy bucket and rope, which I gave them the same evening. And the water in the well is beautifully clear. I bought this bucket and the rope thanks to a money order I had just received from my mother. As it came to under Rs 80, the limit for cash delivery by the postman, I did not have to walk the nine-mile round trip to the post office to pick it up in person.

So often when I give the children coconut oil for their hair, they use most of it on their arms and legs and little remains for the hair. Oh well!

One of the few chickens in our Colony is dedicated to the goddess Kali, because she apparently answered the prayers for good health of a sick woman several months ago. Now this chicken has a roost of branches forming part of a roof covering, and otherwise runs wild! Next May its head is to be chopped off and offered to the goddess.

ENCROACHERS (CONTINUED); HELPFUL CHILDREN

4 December 1966, Sunday

At present I find writing a little awkward, since my right hand is full of Band-Aids. This is the result of taking hold of the rope of an encroacher's cow last Wednesday. Not only doesn't the encroacher like me, his cow doesn't like me either! She just raced down the hill and the rope slid through my hands, as I held on rather tightly. It burned my skin a fair amount and cut into my flesh. It stung!

Anyhow, since at least ten cows were all grazing in a hidden area of the Colony, I felt I was forced to take the step of notifying the Sub-inspector of Police, whom I met in Gudalur two days ago. All this business is terrible! But I have to do it, for there are literally hundreds of cows and buffaloes that "want" to graze on the Colony land.

I was recently reflecting on friends in California (where I studied for my Masters) and their efforts to help in the cause of "Civil Rights". On a different stage, isn't that also what I am trying to do here? i.e. work for the civil rights of these "scheduled tribes" people, this "backward, primitive" hill tribe – terms which rankle, yet are used in documenting the Paniyas, who indeed are no strangers to oppression and indeed slave labour!

I just started a subscription to *The Hindu*, the national English-language newspaper. The bus driver will deliver the paper daily to the tea shop keeper at the bus stop at the corner near the school. At least, I hope so! I feel now a bit more connected to the outside world, reading about Henry Kissinger, Ian Smith etc. I thought it very interesting that in an article on "Motives and Modes of Suicide", the first suggestion, aside from legalistic ones, was for "greater freedom for the younger generation in the choice of marriage partners and greater tolerance on the part of parents in inter-caste and inter-communal marriage." I'm really in a quoting mood today it seems! I'm slowly reading the Hindu religious epic, the *Ramayana*. The wise Rama at one point says: "Time is made up of a succession of tomorrows and one day is just like another. Inevitable sorrows do not become joys by postponement."

Lately both Anthony and I have done quite a bit of sewing on of buttons and mending of torn school clothing. I let some of the children thread the needle, boys and girls alike. They do it without any trouble, even the little ones! Then I let Narayini, the oldest of the school girls, sew on two buttons. She had watched me do it several times. She did it entirely by herself, aside from an occasional stitch she needed my help with. I think I was at least as thrilled as she was!

Collecting money to help here at the Colony is possible. It can be sent to me, but it takes a fair while, since it passes from the Madras[8] main post office down to eventually mine here. But even quite small sums can be so helpful, e.g. C$1 yields Rs 7.50 - enough for about two pairs of children's sandals, or one umbrella.

THE VACCINATOR COMES! "MOTHER HEN" STICKS; PANIYA INGENUITY

5 December 1966, Monday

Without delay I want to write about an incident that happened here today. This morning before the children had to go to school, I hurried

[8] The city name was changed to *Chennai* in August 1996

to the mailbox near the school, at the end of Silver Jubilee Road, to post two letters. On my way back I met Nunjan, one of the oldest of our Paniya school boys – all alone! – on his way to school. He talked to me quite excitedly in Malayalam (Paniya version, of course), but I managed to understand that the rest of the children had run away. He was the only one going to school! Naturally I was curious! With Anthony's help as interpreter, I was able to ascertain that the government vaccinator (for smallpox) and his helpers were about, and had been spied by the Paniyas. Vella, the "chief," had suggested to the children, who at the time were all on their way to school, that it would be wiser for them to run into the jungle to hide. Well, there was still some time before school started, so I decided to try my luck in finding the children. I asked Nunjan to please go on to school, so he could tell the headmaster why the others were all absent. Coming to the Colony houses, Anthony asked some of the women where the children were. They pointed into the distance, where a few of the boys were collecting firewood. At least they were making themselves useful while not in school! For the rest, a vague pointing to the hills and underbrush! I hurried over to the boys and asked them to come to school, still hoping to get the children to school on time by 10:00 a.m.! Slowly some of them dropped their firewood and came to me. I asked them to be my guides for finding the rest. So up the hill we walked. No sign! No sound! The boys looked and shouted. No luck! We continued to some quite dense underbrush; the boys shouted and shouted. Finally a reply from the midst of the thorny shrubbery. My little guides talked and talked, mentioning something about "Durra" being there and wanting them to come to school! A few of the closer ones popped up and together we crawled, literally, on hands and knees, through dense thorny shrub to where some of the rest were hiding. They were standing there, looking at me and laughing at the fact that I was crawling in the thicket! I assured them that I would go with them to the school. For some reason or other this helped, and our "procession" started down the hill. Almost at the Colony the procession halted. Oh, terror! The Vaccinator was just entering the Colony! Some children wanted to turn back. I called their names and they remained "brave". When they reached the houses, some stopped while others ran for it into some of the houses. Meanwhile, I ran ahead addressing the vaccinator rather indignantly, poor fellow!: "See what you're doing! You're causing these children to run into the jungle to hide and thus miss school." By now it was already school time! And I felt that I had worked so hard in getting these 18 children back! The vaccinator was so surprised that I was talking about school children. Wasn't this a Paniya Colony? People aren't accustomed to the fact that so many Paniya children do in fact go to school. Anyhow, I suggested to the vaccinator that he come to my hut after school is out and I would see about helping him then. He enquired where the school was. I didn't like that. He thought he could perhaps get them so conveniently there! But since he didn't have any kind of pain killers, I wanted at least to have a chance to give them APC (aspirin-containing) tablets. He obliged. I didn't really give him any choice, in my heavy-handed manner! So on we marched to school. Some of the children asked what time it was. 10:30 a.m. They actually began to run – a sense of duty or responsibility, which really made me happy! Since one of the mothers was gone, there was no one to look after one baby boy, and his two sisters were now on their way to school. I took the little duffer on my hip, in true Indian fashion, and paced the children to school, arriving just before 11:00 a.m. Quietly I marched the children into their classroom. They immediately sat down, and the headmaster and all the teachers met me, laughing "Very wonderful!" Apparently the attendance today was rather low because of the vaccinations. At least the major part of our Paniya delegation was present, though late! Walking back I met many stares that seemed to say: "This strange Saibu! Running to school with these Paniya children long after school started! And now, with a little Paniya boy on his hip!" One man asked me in Malayalam if it were mine. He looked most puzzled! Since I was in a hurry and couldn't explain in Malayalam anyway, I simply nodded as I rushed by!

Well, in the evening the vaccinator did come to my hut. I had asked Anthony to buy candies, as consolation for the "wounded" children. They wanted me to call the children for candies, then they would have them there to vaccinate! What nonsense! I told the vaccinator that I didn't want to lose the trust of the children in that way! I asked the vaccinator and his men to come to the Colony with me, which by this time was well over half-vacated! I called the first boy, Nunjan, and explained to him: first the APC, then the vaccination, then the "muttai" (candy). He was a little dubious, but willing. It didn't hurt him! Hurray! The result was that half the school children were vaccinated right then and there. This was about a third of the people in the Colony vaccinated today. Some put up quite a resistance. I also insisted that the vaccinators explain what they were doing. They at first said: "These are primitive people; they don't understand."

What made me most happy, even more than the fact that the children present let themselves be vaccinated, was the fact that the children came out of hiding when I went looking for them! It seems I always write something about the children. CUSO-Delhi wrote me that the name "Mother Hen" is mine for keeps now! Apparently they have been hearing stories via the "grapevine", as they put it. Actually they mean the "jungle telegraph"!

9 December 1966, Friday

Today I received a letter from our CUSO-Delhi coordinator which opened: "Dear Mother Hen!" Well – that name seems to be getting around in this country! Also they have sent me an extra medical kit, without my even requesting it. Our Delhi Coordinator said that he has decided to up my monthly subsidy to Rs 50. That, together with NAWA's Rs35 brings my monthly living allowance from Rs75 to Rs85. I think it's very kind of him!

For Christmas I am planning to accept the formal invitation I got yesterday from Art and Flo McHarg, Canadian missionaries stationed on the plains, on the other side of the Nilgiris. But I'm not at all sure yet about when exactly I'll be able to go. Their Christmas Dinner is on the 27th and they want me to come on the 24th. But that means I'd have to leave here the morning of the 23rd and not be back till the evening of the 29th (i.e. away seven days!). I wouldn't be able to do that!

Almost all of the school children on our road now greet me with "Good morning Sir", folding their hands in front of their faces. When I pass them, going in the same direction, they run after me. I now understand the ability of the "Pied Piper" to attract children! I'm almost tempted to practice my recorder more and take it along some time. I'm getting to know several of the non-tribal children by name too. Two young boys, one in elementary school and the other in high school in Erumad, who both know some English, have become two little friends of mine! One was my little helper on the Paniya children's first day of school. I had asked him to tell the other school children to be kind to these rag-clad Paniya children. Often these two boys visit me in the evenings and I say specific words and ask them to translate these for me, which I then copy into my vocabulary list. They asked me if I liked tapioca. Four times already they have brought a good-sized bag full of tapioca. The last time they had it all ready to go, but had forgotten to bring it. Soon their father came, to bring it to me! And it tastes really superb the way Anthony prepares it with fried onions and chili peppers!

About two weeks ago, all the school children were divided into groups according to where they live, and a boy and girl were picked as leaders for the trip to school. Thus, to and from school they line up, single file, one line for boys, one line for girls, beside each other. It's so cute to see them filing by along the road. When they notice me, they all stop and wait till I'm closer, and then almost all at once greet me with: "Good morning, Sir" (or: "Good evening, Sir").

HEAD LOADS; A COOPERATIVE NEIGHBOUR; THREE-MONTH PARCEL WAIT

The other day, coming from Cherambadi with our weekly supply of flour and jaggery (solidified syrup from sugar cane in chunks), I decided to ask (as I had often had the itch to do) the Paniya man who was carrying the flour sack to give it to me for awhile. He resolutely stated: "venda!" ("I don't want to!"). So, through Anthony, I told him that I wanted to find out what kind of a load I was asking him to carry. Reluctantly he let me have the bag, which I placed on my shoulder and carried for about a quarter of a mile. Since almost everything is carried on the head here, my carrying the bag on my shoulder caused considerable amusement. In the evening at paying time, I heard the two men who had been with me discussing how I had carried the bag.

Oh yes, I've started paying only every second day, but some men will have to get used to it. For most it seems to work alright, as far as I can tell after a week of doing it.

An observation I've made concerning the Paniyas: whenever I give them their daily vitamin tablets almost all who have some head cloth, to hold their hair in, take it off before accepting the pill, and then put the covering back on after taking it. While I don't know the basis for this, I definitely think that it is an indication of having culture and traditions!

To prevent one neighbour's cows from overgrazing an area of our field, his men and some of the Paniyas together built a trench and soil levee on top of which a fence was planted from the cuttings of lantana. Hopefully they will root and become a "living fence", a common sight here. Since this fence cuts across a common foot path, I asked two of our men to build a small ladder for crossing the fence. A gate could be left open for cattle to enter. Bamboo poles were cut out of the nearby brush and slit lengthwise, to be made into good strong rope to hold the rungs. A slice removes most of the woody part. Then a man holds one end with his feet and twists the other to make the "rope". Why waste money? And all home-grown – both the materials and the know-how! I love to watch their ingenuity at work! Their canopy for the Musical Comedy gave me the idea to build a similar structure for the mangoes we put in the trenches – to provide shade until the next monsoons! Today this was completed.

On the subject of Kesavan's silkworms, I have watched them since just after hatching from the tiny pinpoint eggs. Now they're already about one third their final size. At pupation (spinning of the chrysalis) they are about 10,000 times hatching size!

13 December 1966, Tuesday

The way my plans now stand, I will take as much as five days off for Christmas. The 24th and 25th I'll spend in Kotagiri at Miss

Armstrong's place. She knows a fair amount of German, and about German customs. So, she is planning a Christmas Eve celebration just for me. I think that's very kind of her. Thus I felt I just had to accept her invitation. Then I'll be going to the missionaries' home, the McHargs, who live in the plains, on the east side of the Nilgiris. They'll have Christmas Dinner on the 27th. Then on the 28th I'll return here. I'm not sure yet whether, with the bus connections, I can make it in one day. I'm glad I'll be in Kotagiri first, since I know there is an English church service which I want to attend either on Christmas Eve or on the 25th, Christmas Day.

Walking along here on a pitch-black night, barely able to make out the silhouettes of large trees, I see the luminescence of fireflies, blinking on and off, scattered over the expanse of a tree – how very like a Christmas tree with candles. Fireflies in the air gliding by or on branches, and glow-worms on the ground or the low shrubs! It's really most enchanting, especially when the air is perfectly still – no human sounds or those of domestic animals. And I can walk along, undisturbed.

15 December 1966, Thursday

The postman came this evening and brought me two of the book packages my parents had sent me just after I left Canada, over three months ago! What a pleasant pre-Christmas surprise. So, now I can play the advent hymns and Christmas songs on my recorder, with music and words at hand! I am so happy about that! One parcel had to be repackaged because of extensive "feeding" it seems, as if by rats. Also my Bible and a Malayalam text were included. Now I can start a real systematic study of the language! However, while the textbook has arrived, the extra time for studying did not come along with it! I'm not sure why not, as I tend to finish with paying of wages, children, patients, and whatever else may be needed, around 6:30 p.m. Then I usually eat supper. After supper I usually have to spend another hour record keeping, counting the money, and such. After all that, I really don't feel up to studying, or even writing letters!

These days the weather is very comfortable. During the day, it's quite warm, but not sultry as it had been a while ago. In the evenings and mornings it is quite cold. So much so in fact, that the Paniyas, who sleep on their concrete floors wrapped in their cloths, beside a little fire, aren't too enthusiastic about rising early. It's been almost two weeks since the last rain. However, as I had leaves and other organic materials placed around the young mango and lime trees, the ground around them is still quite moist, and so we don't need to irrigate yet. Thank goodness! Irrigation always requires extra energy on my part to rouse the Paniyas early in the morning so that we can finish the watering early and don't need to irrigate during the hottest hours of the day. Better for the young trees and better for the Paniyas!

DIVERSE AGRICULTURAL ACTIVITIES

16 December 1966, Friday

Just a little while ago, 14 "little monsters", our Paniya school children, left me, happy with vitamin tablets, coconut oil, biscuits, and the odd replaced slate. The biscuits are from a shipment of 6.35 kg (almost 14 lb) I received free of charge from the "Meals for Millions Association of India", somehow or other through CUSO. They contain tons of vitamins. So daily I distribute them to the children. Such unexpected gifts are sure pleasant surprises!

Some thoughts on the agricultural part of this Colony, on the day-to-day work for which I was taken on by NAWA: Three days ago I started the men on digging a reservoir, at the suggestion of a student/colleague in Davis, California. While I had some definite plans as to the how, once again I waited to see what the Paniyas would come up with. I was curious to note a small circle of soil being left in the extreme centre. By now this has grown into a clay pillar in the centre as the Paniyas broke the soil for the reservoir. Its purpose is to measure the depth! Rather intriguing, I must admit. I would have thought that one could measure the height perfectly well from the

Digging Reservoir Note: clay pillar to measure depth

sides. On opposite sides of the pit, small stairs have been dug, along the walls so that soil can be carried up. I had been thinking about poles placed at the sides and across and a pulley and rope used. This would have been faster. But also more expensive in capital outlay. The longer time it will take now really doesn't matter since this is a reasonably slack period, and in case there was no work we could still pay the men, to support the Colony.

Some men were recently plowing with their homemade wooden plows, with a bamboo yoke. The metal "knife" at the tip and some rope is all that needed to be bought. But I must admit that they just barely scratch the soil! So they have to go many times over the same area. While plodding along, the ploughmen are constantly humming two notes in sequence, like "Do" (upper note) "So" (lower note). The

Paniya men plowing with cattle

first one is hummed twice, followed by the next one, three times, then again the first one twice, and so on. Do, do, so, so, so, do, do, so, so so. The sounds are nice and so is the scene. The bullocks move slowly in front, some still "green" and not wanting to remain in the hand-made yoke, wanting to break out sideways! We have one iron plow, but the men don't know how to handle it well. The iron plow was used again today on my insistence, and I tried it myself for about a half-hour, and then let the Paniya "chief" try. Much better, but also harder work!

After completing the nursery beds yesterday, several Paniyas and I planted some tomatoes, which had just germinated. In the evening I issued some of the seed to various individuals for their private use. I'm not really planning on selling any of the tomatoes, unless much excess is produced. The crop is all intended for these Paniyas. But they feel that something is more theirs if I give it to them personally. Thus they work without pay and do this outside of working hours.

A few men are clearing and burning the shrub for our small future rice paddy, of about three quarters of an acre. Many are cutting and removing old jungle tree stumps which remain in our mulberry field. Also a few men are starting to clear and make miniature terraces and seedbeds on a slope. Almost all of this Colony land is sloped! These beds are for seeding beans and chili peppers. Ammini, whose husband, Karimban, left her, cuts weeds and places them around the mulberries to conserve moisture. A few weeks ago, I had this done around the mangoes and limes, and now while elsewhere the soil is quite dry, it is still quite moist around these trees. Thus we won't have to water for a while. We haven't had rain for a few weeks and might not have much, or any, for a long time!

Last week the main work involved hilling up the soil in the tapioca rows, creating nice loose soil for the tubers to develop in. It's really quite funny; often I think as though all this work has to be done by me, and then am pleasantly surprised if it gets done faster! But I'm learning to think in terms of 25 to 30 men. Actually, I find that if Kurian thinks something requires only one man, for half a day, I had better place two men on the job to get it finished that day! As I said, I'm learning, but there is much to learn yet!

Meanwhile, our problem concerning the grazing encroachers is by no means solved. Now the police are involved on both sides. A petition to get more grazing land has been issued by the grazers against this "Englishman", this "foreigner", "who makes a fool of himself chasing our cows over the hillsides, and who doesn't at all understand our problems!" It's terrible that in trying to help one group, I antagonize another! I do hope I'm not being too heavy-handed in this matter!

CHRISTMAS AT KOTAGIRI; WELCOME "HOME" TO THE PANIYA COLONY

25 December 1966, Sunday, Christmas Day

Kotagiri

A brief chronicle of the last few days follows:

Last week, I went to meet the Collector of the Nilgiris District (administrative "top dog"). I was thoroughly frustrated with him. Here he was, <u>for the first time ever</u> (i.e. he had had this post for six months) in the Erumad area, only a few miles from our Colony. He said that next time, if he had more time, he would come to look at our Colony! Just before leaving him, in probably ill-concealed disgust (I am not a good diplomat), I did mention our grazing encroachers' problem. He stated that obviously there must be a mistake and that such things shouldn't happen. Well, at least he knows about our problem, so should it grow instead of diminish, the Doctor could contact him directly and more insistently.

The most recent "jungle telegram" I received before leaving the Colony for Christmas was to the effect that delegations of Paniyas and Naikas, about five miles from our Colony, wanted to present some of their grievances to me. I don't have a clue what these may be, nor what they feel I could do, but I'll gladly help them, of course, if I can. I guess I'll see.

The kind Miss Armstrong had planned a Christmas Eve celebration mainly because of me, since that is the German custom. I cut a cypress tree on her land and some branches. I then decorated the tree with a few straw stars, which she had from Germany, and some mandarin oranges and ginger biscuits as well. The ginger biscuits, as well as some ginger bread, were made at her request by her servant, quite beautifully. I added some small wooden angels, some hand-cut tinsel, and loads of white candles. Together we decorated the fireplace, tables, bookshelves and pictures with branches, candles, and Christmas cards. A real mixture of English and German customs. My straw "Julbock" (Swedish Yule goat) and a German table runner were placed on the table. At about 7:00 p.m., when all 12 guests had assembled (Doctor and his family; Mr. Phiroze Gandhy; Meridel Fowler, the CUSO nurse; and a few older ladies), Miss Armstrong and I lit the candles. It looked quite festive! The two of us sang "O Du Froehliche" in German, while everyone stood circling the tree. I then played a song ("Lo How a Rose E'er Blooming") on the recorder, while Miss Armstrong and Miss Anna Ornsholt, an elderly Danish lady, who had once worked for Nehru and Indira Gandhi, sang the words in German. Then followed English and German carols and scrumptious sweets and tons of conversation.

I had planned to attend a midnight service but got there a little late. And it was in Tamil. Quite interesting! A band of violins, drums, and more played very rhythmic music to fill the intervals during the service. This morning, Miss Armstrong's servant, and later also his wife, each brought me a cute small flower bouquet of marigolds, poinsettias, camellias, pansies, roses, azaleas – all blooming at once! – taken from the great variety of flowers that adorned both the inside and outside of the house.

28 December 1966, Wednesday

Back at the Paniya Colony

A swarm of letters greeted me last night when I got home at 5:30 p.m., after an expected 11.5-hour bus trip. Thirteen in all. I could hardly wait to read them. But I had to wait, because about 30 or so of the Paniya children and adults came to my hut. Actually, it's so nice to be welcomed back like this after my longest absence so far! I really had the feeling of "coming home"! I showed them photos which I had taken, and which Bev (my fiancée) had developed in California. Now I could distribute some to the people whose pictures I had taken. One of the first things that struck me when I came home was that Karimban was back with his first wife, Ammini! After some time, I was a little mean and asked all to leave and the children to come back this morning, when I gave them a biscuit-a-day for all the days I was gone. In the meantime, I found out that CUSO requested those biscuits be sent for our own, CUSO volunteers' use, since last year two volunteers had to be returned home because of protein deficiency. I do eat some, and I'm sure CUSO won't mind my giving them to the Paniya children who need them more than I do!

To assist the Paniyas, I have requested, and am now receiving, contributions from family and friends. A postal money order works quite well. In this way I can buy exactly the customary things that are needed here. My father (in Oliver, BC) has already sent C$10 for the

Paniyas, and this is now waiting at the post office for me to pick up. My father's mother, in Germany, is sending DM 30, about 55 rupees, and my mother has also sent C$10 to the Doctor for the Paniyas, a contribution from the Ladies' Aid at the St. John's Lutheran Church in Oliver, BC. My sister, Brita, wrote me that she wants to try to collect in her high-school class. I think it's all so wonderful! I've made a list of <u>needs</u>, but I also want to get non-essentials, such as toys for the children. Toys aren't very abundant here; perhaps such could be sent.

Just as I am writing this, about 15 of the Paniya children marched and danced up to my hut with drums! Now they are "performing" for me!

31 December 1966, Saturday

(The last day, the last hours of the year)

Reflections on the ending of 1966 and the beginning of 1967:

How more wonderful can the old year end and the new one begin! No tumult, no people, just my own thoughts! The night air is so crisp and clear. The stars are shining brightly. The moon moves from behind a few narrow lonesome clouds in the otherwise-clear sky to flood the nightly scene. On some early mornings, between 5:30 a.m. and about 7:00 a.m., a light fog often settles on the dewy countryside. The day breaks with a narrow slit of tinted sky in the east enlarging over the crest of the hills. The fog moves through low-lying valleys, making the trees seem almost to move about, as they emanate from a milky nothingness and reach into the field of vision only to disappear, perhaps only their crowns or one side of branches remaining visible. It gets brighter, the distant mountains become more clearly visible and move closer. And yet, perhaps no human voice is heard, nor any man seen. No hamlet, no lone hut. Just hills and trees and shrubs, a brook, the approaching day. It's so peaceful, so wonderful!

The day before yesterday was a profitable one for me – for the Paniya Colony! First I sold some lemon grass oil for Rs13.50 to a store in Erumad. This was from the plot planted just before I came to the

Colony. This time I decided, although new to this crop, to harvest according to my "instinct" rather than to "wait" till Kurian thought it was ready. So I arranged for adequate sickles and the use of boilers well ahead of time. The oil yield per boiler was 350 gm versus 293 gm for the other plot where I had left the timimg more up to Kurian. I am happy this experiment turned out successfully. I now have a better idea what to look for in connection with the proper harvesting stages of the grass.

I went to the post office where a letter and money order from my father awaited me. This was for Rs 68.80 and is now the second sum I am recording under "donations" for the Colony. The first sum was Rs 50, which the McHargs gave me for Christmas. My father sent the money order on December 8th. It took just three weeks to reach me. Quite a bit faster than three months for the packages. I still don't have one of the packages which was sent at the same time!

Yesterday afternoon I received the first real visitors who came here specifically to see me. And they were complete strangers, at that! All are connected to one mission or another. An elder "Miss" from Germany, around mid-30's, Hilde Uber, is a nurse in Sultan's Battery, which is 10 to 12 miles from here in Kerala. She works for the Swiss organization, Basel Mission, and has her own dispensary there. She also arrived in October. And she speaks considerably more Malayalam than I do! One of her patients from Erumad told her about me and our "Harijan (Untouchable) Colony". She also works with many Paniyas and other hill tribes. Then there were her sister and her husband, visiting from Mangalore, in Mysore State, where he works as an engineer in a technical institute, also through the Basel Mission. There was also Rev. Schirmer, from the Lutheran Church – Missouri Synod, residing in a different village in Kerala, not too far from here, together with his wife. Each told a bit about their own work. A delightful picnic took place at my hut, consisting of the finest cake, fruit, and so on. I showed them around the Colony a bit, and it was generally a very pleasant visit. The Pastor and I plan to visit a government-run tribal village, not too far from here, in Kerala. The Pastor has a car! My guests also left me all the sweets, dates, oranges, and cake which we had not eaten at our picnic. I do have a sweet tooth!

1967

RELIGION; LANDOWNERSHIP AND PANIYAS IN CENSUS; AIRMAIL PAPER AND LOCAL SHOPPING

1 January 1967, Sunday

Reflecting on my situation as a rather impecunious CUSO volunteer: I think I would like to be a millionaire, to be able to give more. Not just "handouts", but also to be able to provide the necessary facilities and to organize things most usefully for the Paniyas.

Without adequate language skills, it is almost impossible for me to find out much about the Paniyas' religion. They do have a prayer ground, which lately has been used by some Malayali Hindus. At the Paniyas' request, the Doctor offered to see the District Collector about reinstating this hollow for the Paniyas. I mainly know a bit from the little relevant reading I did. Paniyas are often classified as Hindus. Theirs is a religion which also reveres inanimate objects, like special rocks, more so than other forms of Hinduism I am aware of. Such a rock is found in the prayer ground in the hollow. These Paniyas don't seem to celebrate the main Hindu festivals of the region, but have other festivals. One statement I came across is: "They believe in devil worship, witchcraft and sorcery...." So far I just don't know how true this is. At our Colony we do have at least one Paniyan who does hypnosis and one man known to do palmistry. This man, Chundan, refused to read my palm, saying that he didn't know how to do that.

The dry season is on its way. The hills are beginning to turn brown from the ripening grass. Soon grazing our Colony bullocks will become a real problem!

As the Paniyas cannot be expected to have the resources any time soon for adequate pest control, I am trying to ensure quite a diversity of crops, both for their own consumption and to serve as cash crops. In order to still be able to plant something, we are using the lower slopes for a number of vegetables. From there it is not too far to dig and obtain water from the rice paddy fields. And I assume the manure from the 10 oxen and two buffaloes will be enough to fertilize the vegetables. We planted tomato seedlings shortly before I left at Christmas. These are coming along nicely in their nursery bed under a shade canopy. Eventually we'll transplant them at a wider spacing and tie them to stakes.

The terracing for our bean and chili field is fun! We began with this two days ago. It takes plenty of my time to plan and measure out the layout for the terraces and to see that the details are correct. Since I have never built a terrace in my life, I did enquire about terrace building and saw several types in practice. I hope ours will be alright. Of course, the work is not perfect. But then, I don't expect it to be. Actually, several of the Paniya men have worked on road-building before. They take particular pride in the finishing touches of a terrace. One and a half terraces are now complete.

While I do want the work to go alright, one problem I am always conscious of is that I don't want my presence to make the Paniyas feel that they have to work hard just because I'm there. Then I would be just another slave driver! But some jobs, like the terraces, do require more of my physical presence. For other jobs, I sometimes leave them quite long stretches of time all to themselves. However, I do know by experience that especially on a hot day, they do rest more than should be necessary. While I am happy if a job gets done quickly and well, perhaps it's strange, but at least in my own mind, I don't at all stress efficiency as far as the Paniyas are concerned. I should really be much more efficient in organizing the work than I am! I don't know if it's possible, but it would be so nice if the Paniyas would get a feeling of togetherness, of being part of the group, part of a family working on their own farm. They come from a number of different hamlets. Perhaps I'm dreaming. But it is this "dream" which motivates quite a few of my decisions, which, however, I cannot convey verbally to the Paniyas. Language! Language! And Kurian just can't understand my reasoning.

While staying at Miss Armstrong's place over Christmas, I did some reading about Paniyas. Thus notes to the 1961 Census of India state: "Though the Paniyas are capable of hard work, they aren't attached to their employers and do not give the maximum output." Then, "a land colonization scheme may perhaps succeed if it is properly formulated to suit their temperament." An old article led me to this incredible quote dealing with "domestication" of the Paniyas (what an idea!): "They were savage tribes living in caves and thick forests, coming out only at nights and feeding on paddy and other crops. The Gounden landowners, finding their crops always destroyed by these black beasts, managed to secure a number of them by means of nets known in the Wynad as Thandati and in six months taught them the language and to go on errands. These domesticated Paniyas induced their friends of the forest to join them, and gradually they spread from place to place...."

Considering the future schooling of the Paniya children, I think that having them attend the residential school starting next school year would be the best solution. NAWA does not have enough financial resources for the school children's needs. The small donations I receive from generous relatives and friends overseas are too sporadic for continuous school program needs. And as long as I am here, with my many other responsibilities, I just cannot expect to stay on top of all the needs of the school children. Even just repairing torn clothing is more than Anthony and I can cope with. Many children thus do run around in partially torn clothes, though they do bring the clothes to me once they are aware of a tear.

A while ago I mentioned Karimban K. being back with his first wife. But now I noticed that his second wife is also with him. So now he has two wives. As far as I know, Hindus are only allowed one wife. Perhaps this is an indication of the special religion of the Paniyas, but I don't really know. Jokingly I asked Karimban if there were enough room for the three of them in his small house in the Colony. He jumped up and gleefully replied, most emphatically moving his head "Madhi! Madhi! Madhi!" (Enough! Enough! Enough!).

My relatives may have wondered why I mainly use aerograms and not airmail paper in envelopes when I write. My "saving" is accom-

Karimban with his two wives and child

plished in part by using the aerograms, with 55 nP postage, in contrast to Rs 1.90 for the cheapest, two-page letter in an envelope. I also make my handwriting very small. Anyhow, no airmail paper is available in the two nearest villages, Cherambadi and Erumad. I also found another really quite unexpected problem. Once I had an airmail letter to send, the local Postmaster could not find out what rate to charge for the postage. After all, I am the only foreigner in his jurisdiction. Thus I couldn't send the letter then and there!

In the local towns, used newspaper is generally used to wrap items in, rather than using a separate bag for every small item, as is common at home (in Canada). In many stores, food items are simply wrapped in banana leaves and tied with string. Cooking oil is poured into the buyer's own bottle. To purchase flour, one needs to bring a sack. Many items are just loosely poured into a shopping bag, brought by the buyer.

CHUNDA; ENCROACHER RAMIFICATIONS

3 January 1967, Tuesday

The Christmas Holidays were over yesterday for our Paniya school children. On the first day of school in 1967, our Paniya children could march along to school with their newly acquired shoulder bags slung over their shoulders, with the slates inside. The Rs 50 Christmas present I received from the McHargs went for these items. Also yesterday the children could finally participate in receiving the school's midday meal. The government subsidizes these meals, and we (NAWA) need to pay just Re 1 per child per month. This is a wholesome meal, and even includes a drink of milk in the afternoon! While yesterday they ate their meals on banana leaves (as is a most common practice here), during the afternoon I went with Anthony to Cherambadi to buy enamel bowls for the children. These bowls were purchased with funds sent by my father, by postal money order (PMO). I had enough money left to also buy one package of 24 small soap pieces and 10 combs, enough for half the school children. As these PMOs can be delivered to me by the mailman if they are under Rs 80, it is quite convenient for me to receive them in units of C$10 (which is Rs 68.80 now).

I've mentioned that Karimban now lives with his two wives. Well, a couple of days later another wife of the "chief", Vella (Karimban's brother), appeared. Vella was living with his third wife, his first one having died a while ago. So, now he lives with both his second and third wives.

Another member of this family, their sister, Chunda, has been causing some trouble lately. Before Christmas, on the last Doctor's visit, numerous men appeared with part of their midday meals (prepared by Chunda), complaining that they weren't boiled well enough. This Midday meal is provided by funds from NAWA. At once, Kurian scolded Chunda, in the presence of the Doctor and all the men who appeared. No attempt had been made to determine why the food wasn't boiled enough. It appears that Chunda had not been provided with enough firewood to properly boil the food! But by now Kurian had humiliated Chunda in the Doctor's presence, and since I was leaving for Christmas two days later, with Anthony leaving the following day, she wasn't going to work as long as Kurian was in charge. She went on strike and didn't cook! So one of the men cooked, and apparently did quite a good job. On returning, I asked Chunda if she would again cook, and now she does it again. But there have been other troubles as well. Apparently in quite scolding tones, so I had been informed, Chunda told members of about 20 of the Paniya families which joined the Colony, that they came essentially as parasites, to take what was originally intended for her and her kinfolk. The five families which originally lived on the Colony land and whose hamlet had burnt down, were the first ones approached by the Doctor about setting up this Colony. It was Chunda who had at that time (predating my presence on the scene) convinced the other Paniyas to trust the Doctor with treating their sick. This former hamlet was called Karikkamunda. Well, some of the scolded Paniyas are quite sensitive and thus refused to eat food cooked by Chunda. All isn't well here yet. I think Chunda should apologize. I'm afraid if such outbursts are repeated, I will have to oust her as cook, for the benefit of the rest of the Colony. One day only half the people ate the midday meal! Of course I wouldn't let Chunda starve, and would find various odd jobs for her to do. As she is a widow, she has no male "provider". But I really don't want to fire her as cook!

Yesterday, when Doctor & Co. visited, I decided to officially join NAWA by paying the yearly dues of Rs 5. The two ladies, Miss Armstrong and Miss Martin, told me that I am just the third Caucasian to join the Association, they being the first two.

There have been some repercussions stemming from our encroacher problem, which otherwise has been latent for the last few weeks. The Doctor had become very annoyed about the complaint letter sent to the police, especially since in it they mentioned this "foreigner", who doesn't at all understand them and who tries to make this land his own. The Doctor mentioned to various officials we visited together before Christmas in Ooty, and in his letter to the Central Government in New Delhi, that here we have a volunteer who comes to help these Paniyas, and he is branded a "foreigner" in the same category as probably "colonials" are branded. Anyhow, Dr. Narasimhan wrote to R. Diraviam, the Director of Rural Development and Joint Secretary in the Government in Delhi, who is coordinator for Madras concerning "Foreign Volunteers". Diraviam immediately issued a notice to all the Collectors in this Madras State, requesting information as to which volunteers work in their respective Districts, for which agency or association they work, and by which agency abroad they are sponsored. This directive mentioned, as an example, the CUSO agriculturalist and the nurse working in the Nilgiris District for NAWA. Our Nilgiris Collector relayed the directive to the various Block Development Officers (BDO's). The BDO from the Gudalur Taluk then sent a man out to me and asked me to fill out such a questionnaire. Thus the circle was completed! While I dutifully filled out the questionnaire, only yesterday did the Doctor tell me that this was in direct response to his action upon my informing him of the content of the complaint letter against me by the encroachers. I do so hope that this "encroacher problem" will remain latent and eventually die down entirely!

PERCEPTUAL ACTIVITY; POISONOUS PLANTS

5 January 1967, Thursday

The following are some comments on a section on perceptual activity in children from a Russian psychology book, which I am still trying to send as a present to Bev. She is studying psychology in California, and Russian books are not readily available in the US. Parts I found quite interesting, specially relating to the methods used. Other parts were more or less as I would have expected. It would be rather interesting to see how the perceptual activity would compare to a group such as the Paniya children, where quite different aspects of perceptual processes may be early internalized, while at the same time perhaps recognition and discrimination for certain characteristics which an urbanized child may acquire early in his childhood, may not be at all, or only very poorly developed in a Paniyan. And for comparison/contrast, what urban child can go along the road and woods and tell you which plants there are edible and which are deathly poisonous, as even tiny tots among the Paniyas can! Indeed, recently after the school children were at school, several of the tiny tots walked along with me and offered me some leaves and berries to taste, from local shrubs. I looked at these and slowly raised them to my mouth. The one little duffer slapped my hand down and told me emphatically, "Moriche poy! moriche poy!", indicating that I would die if I ate these! And then everyone, including me, laughed heartily!

Anyhow, I am wondering whether or not the same processes or lack in their development is involved in the inability, on first trial, of some of the Paniyas to form even vaguely similar replicas of their names in large block letters. I tried this with quite a few of the Paniya men. Some just very slowly drew most scraggly lines and others placed letters all over the place with no regard for linear direction. Of course, other factors may and no doubt are also involved, such as coordination of fingers and eyes. But still, even these are related to the discriminatory processes involved in the individual's development of perceptual activity.

FORMAL FUNCTIONS AND POLITICIANS

8 January 1967, Sunday

Ooty, 10:30 p.m.

An interesting day of officialdom to reflect on.Two days ago I received a letter from NAWA indicating that I should travel to Ooty to attend the official opening of an Indo-German agricultural project, but with no details of when and where. Yesterday I walked the four and a half miles one way to the post office in Erumad in the hope of another letter with further information. Alas, no letter. That evening one of the Panichis (Paniya women) brought me a telegram, which the postal runner had given to her when he saw her washing some of the children's clothes at one of the two Colony wells, early in the afternoon!

The "inauguration ceremony" for the "Indo-German Project for the Nilgiris", which I attended with the Doctor, was quite interesting! The speeches were clear and to the point. The German Project Leader, Dr. Geuting; the German Ambassador, Baron Dieter von Mirbach; and the Minister of Agriculture and Forestry of the Federal Republic of Germany, all spoke, as did the latter's Indian counterpart, Shri C. Subramaniam, the Union Minister for Food, Agriculture, Community Development and Cooperation. The stage was decorated generously with flowers and garlands, plus a floral gate at the entrance, with a red carpet. While I had arrived before noon, well ahead of the program, the Doctor and entourage showed up quite late. Nevertheless, we were all ushered in through a side door to sit right in the front row. It turned out that the project would concentrate heavily on potato production for the Nilgiris District, but our Colony is no longer in the potato-growing region. It's generally too hot at our much lower altitude. Nevertheless, other vegetables and fruits, as well as agricultural implements, are to be dealt with in the project. Well, I do hope we'll get something out of this. The Doctor is sure trying hard! It's fascinating to watch him operate. Somehow after the ceremony he managed to make the rounds of all the politicians who were present. And there were a lot, as elections are scheduled for February. The Doctor managed to get an interview with the Union Minister for himself, me and the two "European" ladies! All the other people were chased out of the room, the doors were closed, and a police guard posted outside the doors! The Doctor pushed me forward, introducing me as an agricultural volunteer who had come to the Paniya Colony. The Minister asked us to be seated, asking me to sit in an easy chair next to him, and for about 15 minutes the Doctor spoke mainly about the work and aims of the Paniya Colony, providing details of my combing and oiling the children's hair, sending them to school, and also adding some embellishments. Once that interview was over, the Minister for Food and Agriculture for Madras State, Sri Kakkam, had his turn with us. The Doctor and Sri Kakkam had known each other since they were young lads. The Doctor captured him just as he was about to leave. Here too, the Doctor said: "Here is Henning, agricultural volunteer representing the Canadian University Service Overseas..." And then the two ladies were duly introduced. Somehow my upbringing caused me to feel uncomfortable when I was introduced before the ladies were. But then, this meeting was in connection with agriculture. So, although the itinerary arrangers for the Minister noted that he was already behind schedule, the Minister bade us sit down, asking me to sit next to him on the bed in this Government Guest House. Again the Doctor outlined our work at the Paniya Colony. Having received the Minister's word that he would do his level best to get us included in the Indo-German Project, we took our leave with cordial handshakes. Well, let's wait and see what happens! By now, many of the local politicians, both State and Federal in the neighbouring regions, were urging the Doctor to have them introduced to us three "Europeans". After all, the two Ministers, despite their rushed time, had granted us interviews. Yes indeed, though actually quite shy publicly, the Doctor is amazing to watch in what he will do for the "cause". It's so fascinating to watch his tactics.

Just yesterday, I've been told, the Prime Minister, Mrs. Gandhi, came to Ooty by helicopter to address a public meeting and then inaugurate the Hindustan Photo Films factory there. As it happened, the Doctor had her written word from a few months ago that when she came to this area, she would try to talk to him in person. Assuming that Miss Ornsholt's presence would ensure a meeting with Mrs. Gandhi, the Doctor yesterday handed her over to a federal Minister, to be included in the honorary car cavalcade. Miss Ornsholt is the elderly Danish lady who lives in Kotagiri and had worked in the Nehru household years ago. Then the Doctor arranged for a whole group of Toda women to present Mrs. Gandhi with a bouquet of flowers. They lined the street on her way back to the helicopter. Mrs. Gandhi saw the Doctor and the Toda women and ordered her cavalcade to stop, though she was already over two hours behind schedule. At first the cavalcade did not stop but then Mrs. Gandhi almost shouted at them to halt. She then chatted with the Doctor for a while and the Toda women had an opportunity to present their bouquet. All the while the whole cavalcade waited!

Back to today. It was so funny to see some of the politicians in action. Some seriously wanted to engage me in their election campaigns, on my side of the Nilgiri Hills, in my region among the tribes, since I am familiar to at least some of the tribals there! Well, I would not involve myself in such activities, as I am only too aware it would be illegal for me as a foreigner and visitor in India!

11 January 1967, Wednesday

Paniya Colony

Yesterday, I again marched the nine-mile, round trip to the post office, just to check whether any mail had arrived. As I am reasonably cut off from the outside world, mail becomes very important. Delivery to my house is only on Mondays, Wednesdays and Fridays.

I still have not been able to send that book to Bev, in spite of trying at three different post offices. Such poorly run services, whose efficiency we take for granted at home, can really help create "culture shock", especially for someone heavily dependant on such services. I tried at Cherambadi – no custom's declaration forms. I tried in Kotagiri – it was too much bother for the teller. And I tried in Ooty – they could not find the proper gift parcel rate by the time my bus left for me to get back home (to the Paniya Colony).

Meanwhile the situation with Meridel, the Canadian CUSO nurse, is very confusing. She may not end up coming here after all. However, I do know that I definitely need help, and the Doctor realizes this.

MY DAY OFF??; MONITORING BUILDING OF HOUSE; VISITORS; PANIYAN LANGUAGE

15 January 1967, Sunday

It is encouraging to receive support in print, even though indirect, for some of the agricultural reasons I have given for deeming it essential to keep the grazing encroachers off the Colony land. Thus in today's *The Hindu* (proclaimed as "India's national newspaper"), in the section titled "For the Farmer's Notebook", under the title "Four fodder finds for hill areas" it states:

> Luxurious as the vegetation in many of the hill slopes in regions like Nilgiris or Kodaikanal appears, it is sometimes of little use to either man or cattle. Whatever grass or other species of rich fodder value grows there, is often overgrazed by the cattle and sheep allowed to forage in the area.
>
> The lacuna created by this destruction gives room for the spread of other species of poor utility like ferns or rough grasses. The whole tract becomes often unfit to sustain quality livestock and at the best can support inferior stock with meager productive ability.

My subscription to *The Hindu* is being delivered, as planned, to the tea shop at the corner (Kayyunni) of our Silver Jubilee Road and the main road to Erumad and Cherambadi, about one mile from my hut. The name listed as the recipient says simply "Manager, Konnachal". That refers to me and my being Manager of the Paniya Colony in the postal region of Konnachal. Delivery to this tea shop is to be daily. In January, so far, about seven issues have not arrived. Oh well! More or less "daily"! The inauguration of the Indo-German agricultural project which I attended in Ooty on January 8th, made front-page headlines in *The Hindu* last week.

It's Sunday today. My "day off", but in a way not, as I stay right here, near or in my hut and am "on call". The Doctor may come this evening. Anyway, where could I go and find a quiet spot to write or read? This morning we had troubles concerning the Community Kitchen food preparation. Chunda did not show up today! Then, many of the men came to me at odd times requesting wage advances. The subcontractor for the house asked about money, just in case the contractor didn't show up, as he's been gone for celebrations for the past four days. Then, the odd patients appeared. And numerous children brought clothes for me to fix. The sewing took up most of the morning. Hordes of children in smaller or larger groups flocked to my hut every half hour or so requesting vitamin pills, biscuits (I have run out), slate pencils, and soap. Yesterday was washing day at the well. With several mothers watching from nearby or some distance away, I scrubbed quite a few of the little ones with soap myself, and then poured buckets of water over them. But I also gave the soap to the older children to help wash the younger ones, and they even wash their own clothes now. Then came a request for more balloons. Yesterday I distributed them randomly, but had only 30 while there are over 50 children in the Colony all told. Things like this make it tough to give "small things". With such huge numbers, even these small things become costly. Then I joined some of the children in playing ball. I gave a few balls to those who didn't need slate pencils. Other children came and "performed" for me parts of a Paniya drama which they had attended last night. As it was at some distance from our Colony, I chose not to attend. I enjoy playing with the children, but even this can be all-consuming.

Over the last few days I have often gone over to the new house being built. Continuously I see minor or maybe major items which to me seem strange or incorrect. One side of the roof is four inches lower than the other, over the verandah. But what do I know of house building! There are just too many things I notice which seem odd. But then, how many other items are there which may be incorrect which I don't notice? I need to keep checking on the work all the time. When the Doctor made some suggestions for changes, the contractor did not write these down. So, days later he asked me what those changes were. As though I had the contract! Occasionally I have suggested some changes. And the construction is to keep going. Sometimes the contractor has asked for my interpretation of items on the building plan! Aside from Grade 9 Drafting, I have zero experience in this field. Although the contractor again flew the coup for a few days, the subcontractor and his helpers – a man and five small boys – worked quite quickly. This man is a mason and they require their payment. It was fun watching the brick masonry in operation, especially the laying of the vaulted structure over the doorway with the keystone in the middle. I found it also really interesting to watch the clayey mud being applied as "glue" to stick on the auxiliary bricks along the walls. Today is in fact the 15th, the date the house was to have been all completed according to the "revised" schedule which the contractor still maintained as being in effect as recently as three weeks ago. Will I ever be glad when this house is completed!

Right now, at this minute, there are distant rumblings of thunder in the mountains. So – who knows? – some of it may come this way. It would be the first thunderstorm in two months. A few days ago we had the first rain for over a month. While it wasn't a lot, it was enough to permit us not to have to water the limes and mangoes this past week. The rain also helped the beans, which we had planted on the three terraces. They have now nicely sprouted.

On January 10th, I had a visit from Olav Liengaard, the Danish Volunteer Service Coordinator for India, and his wife. They were really interested in this Colony. I showed them part of the field work and also the houses. A while ago I had Anthony buy some rope and nails and put these up as clothes lines, in each house, so that cloth and other items would not have to lie on the ground. This was still from the money donated by the McHargs at Christmas. Anyhow, seeing some of the drums hanging over these lines, I thought of asking the Paniyas to play for the visitors. I beckoned the flutist, who approached us. He indicated that he didn't have his flute. I felt his clothing and lo and behold found the flute hidden there. He is a cheerful chap and likes to pretend not to have something only to be amused at my searching and finding it.

Several of the men played and Olav, who loves diverse music, retrieved a tape recorder from his car to record the music. Well, what a surprise on the Paniyas' faces when Olav played the music back to them! The flutist's mouth dropped open. And one of the women just ran up to the machine to watch more closely. Then it went as follows:

some playing; then playback and repeat. Time and again some of the little children came up to me stating they wanted the machine to "sing" again. It was so cute. The atmosphere was great, so I thought why not ask the women to dance their circle dance. And they did! Even without any payment, which the Panichis first laughingly requested of me. I was so happy! One woman even agreed to sing for the tape recorder. Now I know that I can get the Paniyas to play their music and dance on request! Until now I didn't know if I could ever get the Paniyas to perform during the daylight hours. Perhaps there may be an occasion for a photo yet!

I have just moved outside my hut to continue writing. It was getting too dark inside as it is rather overcast and I don't have any windows, just two planks which serve as my door. My "chair" out here on the veranda is the box in which the 6.25 kg of biscuits came. That's the only chair I now have. As Kurian moved the rickety old chair, and the even more rickety table to where he is staying at the Colony, that is where I now go to pay the wages. This suits me better than paying from my veranda. I don't like being watched by oodles of eyes as I unlock my trunk to get the money out. Yesterday as I was paying there, Anthony was counting for me, in English, how many men desired a wage advance, "one, two, three". Then Vella piped up, "four, five, six", and laughed sheepishly as though he was quite astonished at what had come out of his mouth. It was quite funny! But I do think every one of the children here now can count to ten in English. As they count out loud so often, it is really little wonder that the men and women pick it up. Similarly with "good morning", "good afternoon", and "good evening".

16 January 1967, Monday

This was a really quite busy, but fun morning for me! Anthony was still away, as his "day off" is from Sunday to Monday just before lunch. Kurian has also been off since last night in a neighbouring town. However, no word was mentioned to me about this – somewhat annoying. I instructed Vella on the work planned for the men for the day. Then I had to provide the school children with odds and ends and send them off to school. I think such attention helps to improve their school attendance. By contrast, one day when I had to go to Erumad early in the morning, and thus could not attend to the children, only ten attended school. That's quite terrible! I also had to try to keep the little children from crowding too closely and eating the tapioca raw while the women were peeling it. After all, that reduces the allotment for the midday meal. This has led to much quarreling lately.

This morning it was a bit cloudy and not too windy, ideal for spraying some of the pesticides I had made ready for the mangoes and limes. I do the spraying myself, but about four of the Paniya men bring the water and carry the sprayer. After having many (paid) volunteers one day, the next day there were none, although they could rest quite a bit in between carrying, while I chased up and down the hillsides with the hose from tree to tree all morning long. Oh well!

At the work site for the men, the chief's brother, Urali, was telling me all kinds of things, but of course in their "Paniyan Basha" (Paniya language, closely related to Malayalam), interspersed with such English words as: no, yes, school, today, tomorrow, buffaloes, cow, right, tapioca, lemon grass. It was such fun! Actually most of the morning turned into a real two-way language session! The Paniya men got a real kick out of the fact that I constantly took out my little book to write down the word and its translation. I was so pleased that they already knew several English words, most of which, I flatter myself (and don't think I am wrong in this), they have acquired since my coming here.

One of the men asked me to show him my watch. He jumped with glee shouting "12 o'clock!" But another man stated that it was just 11:45 a.m.. These times were told in the Paniyan Basha. I checked and it was indeed 11:45a.m.. This second Paniyan told me he looked at the sun to tell the time. Nonetheless, the one who had looked at my watch had at least been able to tell the time to within fifteen minutes! Imagine that! And at least he is consistent! Just now the same man came to my hut asking for his pay. It's 4:45 p.m.; thus he is 15 minutes early (again)! By the way, telling time is one of the things I try to teach the children. Several of them can do it to within 15 minutes accuracy by now. Perhaps none will ever own a watch, yet it is one of the few things I can teach them in spite of the language barrier! As for the Paniya men, I keep letting Kurian tell them that they'll know English before I'll know Malayalam.

It's evening now. While paying the Paniyas their wages my informal "language lesson" was extended to certain words and phrases which differ quite a bit in the Paniyan Basha from Malayalam. If only I could do this kind of study more regularly and more often!

It is really so pleasing, to me at least, to watch the Paniya men, after working in the clayey soil all morning, take turns washing their hands, lower legs, and feet before going to take their midday meals! The Doctor and Miss Armstrong, on their most recent visit, commented on how the Paniyas did look cleaner and their skin brighter now. Miss Armstrong stated furthermore that she did not think this was an accident. Perhaps not. After all, during the past week I distributed 20 bars of soap to the children and a few of the men. While a lot of the soap is wasted, I am sure the parents get to use some, too. Actually it makes me quite happy that quite often now several of the children come asking for soap. But they really do have to learn to share better! Especially some! Kids will be kids, I assume.

So many times I have watched several of the Paniyas pick one of the flowering roses in front of my hut, as do I, to place in front of Bev's picture. The women occasionally put a rose in their hair. Several times I've seen one of the men pick a rose and hand it to his wife or their small child. They love flowers! It's so neat, this international language of beauty. One day, when Chunda came to the nearby roses and placed one in her hair, she swayed back and forth, preening in front of me, as I sat on my veranda trying to write. I really did not know what to think! Her aged mother spoke to her a bit gruffly, it seemed, but of course, I could not understand the words. Anyhow, I must admit I felt a bit embarrassed and tried to ignore what had occurred there in front of me.

The other day, Elukan, one of the older Paniya school boys, pointed to two stick people drawn on their house wall and exclaimed, "Adhe Durra; adhe America" (this is "Master"; this is America). I thought it was cute. He had seen the photo of Bev and me in my room, and I had Anthony explain that Bev was 10,000 miles away in America. This Elukan is a real teenager! The other day I was writing names on the children's palms. He promptly asked me to write his name and that of one of the girls on his palm! The girl protested! On another occasion, when Miss Armstrong gave someone a photo of a group of the Panichis and girls, Elukan ended up with it, pointing to one of the girls, Onathi, and stating that she was his! It is true that the Paniyas marry quite young. Several of the young men who are now married are still likely in their teens (the average age of marriage according to Indian statistics in the decade from 1951 to 1961). Most recently, two of the young men had married by elopement.

Elukan and Nunjan on Silver Jubilee Road

DEATH IN THE COLONY; THE KALI FESTIVAL

18 January 1967, Wednesday

During the night, the third wife of the "chief" Vella died. He had lived with her since well before the Colony started. His second wife had joined him recently again. The flies were crawling all over the dead woman's body, despite its being indoors and surrounded by 10 people. I heard of the death by messenger. I was six miles away at a temple celebration of the Paniyas and Chettis. The news reached me at 1:30 a.m. I don't think that I have ever walked six miles at such a fast pace with no stops and without tiring. When I reached the Colony, I took a kerosene lamp from my hut since it was pitch-dark and my flash-light batteries had run out during my walk. Anthony and I went to Vella's house, where he was squatting next to his dead wife and sobbing. I simply put my arm around him and pressed him tight. Yes, tears also came to my eyes. I won't forget this death. After all, it was the first time in my life that I had touched a cold, dead body. Also I had known her, and a short time earlier she had been alive.

Today I cancelled all work. I still paid the Paniyas half a day's wages as I couldn't let them starve. Food had to be cooked. Although one of the chief's brothers stated that no one would eat, a bit later others came and requested a meal. I arranged for food to be prepared, even though late, and it wasn't eaten until after the funeral, around 4:30 p.m. or 5:00 p.m. In the morning the men had spread out in all directions to notify relatives. About ten of the women stayed together with the body, in one room. While many Paniyas, mostly relatives from the area, came, only the men went to the grave with the dead body.

Saturday is to be a "pujah" (prayer) festival. From what I can gather, this is to be a cheerful celebration. But Vella is understandably far from cheerful now. Such a situation! If only I had been at the Colony at the time; if only I had had a vehicle I could have sought out the tea estate doctor! This death really emphasizes my relative immobility here. I know it is too late now. But for the future!

This morning I went early to the post and telegraph office to notify the Doctor by telegram. I then enquired about the whereabouts of the estate doctor and even managed to meet his wife. She is also a doctor. So, I now have contacts for future emergencies. All along I have been afraid of sudden deaths in the Colony! How will it affect the Paniya's attitude to the Doctor and to the rest of us? Will any gained trust be put to naught? I really don't know. But I so pray that God will also guide and help us in this potential crisis.

The Doctor had been treating the woman. On his last visit he said that he'd rather not take her to one of his hospitals just then. She was too sick for the long jeep ride. He gave her pills to reduce her swellings. He hoped to be able to transport her in the jeep in another week or so. Over the past few months, ever since I have been here, her body and face have really swollen up from this anemia.

On my walk from the temple to the Colony, I was constantly wondering whether the child she was carrying, in its sixth month, could have been saved if a doctor had been at hand. Would it be viable? I wasn't here at the Colony when she died! I could neither get to a doctor nor did I even know how to find a doctor in the middle of the night! Today the doctor at the estate told me that with the existing facilities (no incubator!), the child could not have been saved since it was so premature. Since leaving for the temple yesterday evening at 7:00 p.m. until this morning at 10:00 a.m., I have walked about 20 miles! I do need a means of transport for emergencies, and in general. The burden of responsibility for the failure and success of this Colony really does weigh heavily on me!

Yesterday evening's visit to the temple celebrations, in contrast to events later that night, was so very cheerful! I walked along with a whole group of Paniyas from our Colony, about 25 or 30, carrying one, sometimes two small children who expressed their delight in wide smiles. At the temple, under the open sky in a forest glen, people all sat in a semicircle on a slope, waiting for the parade. I went with our group to sit on the Paniya part of the hill. Our boys pushed and shoved to sit or lie next to me. Yes, it was a long wait. One of the boys had worn my coat and I his cloth, slung in Paniya style around my shoulders, for the whole walk there.

Later some of the Paniya men asked me to join them, and several of the boys asked me if they could come along. After ending up buying tea and eats for 19 Paniyas, I sat on the ground, covered by one of the cloths, between two of the Paniya men, with the boys around my legs. When the parade finally came, I alternated in putting the smaller boys on my shoulders so they could see better. Oh yes, while we were all waiting, several times such a crowd gathered around us! Pushing and shoving! Also to "see better", but in this case, people wanted to see the Paniyas with their "sahib". One of the teenaged boys, who was not a Paniya, and whom I did not know, told me in English that they are all looking because they like me very much because of my affection for the Paniyas. Later one of the men asked if he could get a job in my "department". But I'm just too skeptical about this lip service of interest in the tribals! The trouble is that for many it really is an expression of status to be talking to this "sahib". I was the only "European" there. Of course curiosity is a human trait. But I always disliked this "brown-nosing", trying to get into somebody's, anybody's, good books. Yes it is frightening to me what power white skin still seems to have for so many people here! I also realize that quite possibly I appear at times too grim towards non-Paniyas here. Even so, at this festival, I mainly joked about with the Paniya boys and men, largely ignoring the "starers".

Then the parade came! There were flocks of people – a whole drummer "squad", many men dressed in a fancy red and gold type of dress as attendants; and one man, as the goddess Kali, whose temple this was. Then came women carrying candles and flags and standards, followed by a very ornately, nicely decorated elephant with three men plus parasol on top. A most significant feature of this parade was the model of a cow. It was carried by six men, including one of our Colony Paniyas, and raised and lowered to constant shouts. Somehow this part seemed a bit eerie to me, as it reminded me of the biblical story of Moses going up the mountain to receive the 10 Commandments, and how the people made a golden calf and danced about it. The moon was still up, providing light, as did diverse lanterns hung all about the place. The parade encircled the temple several times to the sounds of singing and instrumental music. Then a fire-juggler performed. While he performed his fascinating feats to the rhythm of drums, a little ways removed, several Paniya men, including some from our Colony, danced to the drums and flutes. At that point one of the Paniyas beckoned to tell me the sad news of the death in the Colony. Naturally, we all prepared to leave, but it would take a while for everyone to find each other in the dark and gather together for the return walk. I decided to put the boys I was with in the care of some of the Paniya adults and hasten away myself.

The anemia which Vella's wife died of claims many lives here. But one of his sons, a stepson of the wife who just died, was cured in two weeks in the care of the Doctor at one of his hospitals in the Nilgiri Hills. His condition had been about as bad as his stepmother's had been four to six weeks earlier. When he came back here, I was so pleasantly surprised at his completely new appearance! He greeted me with 'salam' and extended his hand. I was really touched! By the way, his was one of the elopement marriages I referred to earlier. A real wedding is planned within the Colony in the near future, as both the groom and bride are members of this Colony. The father of the boy, the artist/hypnotist Veran, is the father of two of the girls who have been married since I came here. After this planned wedding, only one of the men on the work crew will be unwed, aside from a widower.

The smallpox vaccinations took place six to seven weeks ago and in the weeks since, quite a few of the children have in fact been sick. It seems to be a direct result of the vaccinations. Those were not easy days! The pain tablets I gave out proved to be useful, as they did indeed help reduce the pain. Some of the children still had a fever three and four weeks after the vaccinations. Other children were very weepy and just not themselves for a considerable time. One day only 10 of the Paniya children attended school. Terrible! But as I just don't have the time often required for the children, I cannot really handle all this, as well as the work load concerning the Colony farm and the adults.

As I have grown a mustache since coming here to the Paniya Colony, I feel I should comment a bit. While some might consider the mustache a symbol of the former colonial masters, I feel quite certain, since the Paniyas did not have European colonial masters, that this would not be an issue with them. Furthermore, the few British "planters" in this area who could be considered a remnant of colonial times, do not have mustaches. Yet almost half of the Indian men here do.

Today is exactly 100 days since I arrived at this Paniya Colony!

20 January 1967, Friday

Today I was finally able to send the Russian psychology book to Bev. It had taken visits to three different post offices, including four visits to the one in Cherambadi. Since the postage cost Rs 30, and only small denomination stamps are available in these small villages, I needed to affix 130 separate stamps!

As Dr. Narasimhan studied in Madras (I think), he only studied "western medicine", but he knows a great deal about Ayurvedic medicine. Since in his practice he has very limited time, he cannot get involved in using it, for it requires that the herbs be collected, boiled, and prepared in specific ways, prior to being rendered into a useable form. But I am fascinated by his knowledge of these herbs. The Doctor knows the Latin names of all these plants and knows what chemical substances are the active ones in the preparations made from the plants. It is certainly not a matter of ascribing mystical powers to the plants! I am trying to persuade him to record all this knowledge, including his treatments. For some diseases, such as Beriberi, these herbal treatments seem to heal more effectively and quickly than the treatments used in modern hospitals. In the near future, the Doctor is expecting an Australian pharmacist volunteer, who presumably will be able to assist in making the preparations, as well as recording the disease histories.

Wild pigs are starting to cause trouble in the tapioca fields. So I let the Paniyas build a little raised hut for a night watchman. This will hopefully also reduce night-time theft!

Urali, who really wants to learn English, recently told me, of course through an interpreter, that if they (the Paniyas) were asleep in their homes and I would call one of them in the night, I would not receive an answer. They would be afraid of being called by a devil.

The good soul, Anthony, who is trilingual (Tamil, Hindi and English) helps me a lot in making changes, and changes again, in the midday meal ingredients and quantities for the Paniyas. Availability in the market, nutritional value and cost of ingredients influence our choices. Anthony does the purchasing of the supplies, he weighs the ingredients out daily, and when he and I decide on such changes, he then demonstrates these to the Paniyas, so that know what they are eating and can follow our instructions for future meal preparations.

22 January 1967, Sunday

Last night, I was asked to come to the home where Vella's Panichi had died during the early hours of Wednesday, to attend a Paniya "pujah" (prayer ritual). There in the dark, outside the home, squatted Vella and his relatives and some of the other Paniyas. Kavalan, with his great shock of hair, was standing completely motionless for quite a while, his legs spread slightly apart. Gradually his whole body began to tremble and shake. After some time, he almost lost his balance, stomped off in one direction for about 30 feet, did an about-face and stomped again. Back and forth he went. He accompanied his stomping with what seemed like an outburst of cut-off, quickly uttered, loudly shouted words or phrases. Apparently he was now possessed by a god. About a quarter hour into this back and forth stomping and shouting loudly, Kavalan picked up a large knife, a cleaver actually, holding it in one hand. After a bit more stomping, he suddenly came to a complete stop, and with one whack of the knife cut a coconut, which was lying on the ground, in two. The "pujahs" were finished. I was the first one to be offered of the sacrificial, sacred food, consisting of the coconut together with a roasted rice preparation.

LIONS DESCENDING AND THE AUSTRALIAN VOLUNTEER

29 January 1967, Sunday

All week long I didn't get a chance to write. So now I am sitting outside on the hillside, having escaped from my hut. This week has been busy!

The day before yesterday I received a Christmas parcel from my grandmother Mündel in Germany, sent November 4th! Everything arrived in perfect condition, a whole bunch of special Christmas sweets. Good timing, too. This week I have a guest. It's the Australian volunteer pharmacist, John Holds, with Community Aid Abroad, who arrived in Kotagiri last weekend. Today is his 26th birthday. At least I had a few choice sweets to offer him. Right now we are both sitting here in a shady spot and writing.

A few days ago an article appeared in the national newspaper, *The Hindu*, covering two columns of the Special Republic Day Supplement, featuring Miss Ornsholt, the elderly Danish lady who had worked in the Nehru and Gandhi households for 28 years and now lives in Kotagiri. At the end of that article there is a reference to Mrs. Gandhi's visit to the Nilgiris and meeting with Dr. Narasimhan, as follows:

> While returning to the Race Course to board the helicopter to return to Coimbatore, Mrs. Gandhi spotted Dr. Narasimhan, the well-known tribal worker in the Nilgiris, standing among the crowd. She stopped the car and asked Dr. Narasimhan to look after Miss Ornsholt, who was left behind at the Films Factory, and take her back safely to Kotagiri in his jeep.

Thursday, the 26th, "Republic Day" in India, was a big day in the Colony. The Doctor and all the volunteers, the two elderly ladies plus about 50 members and their families of the Nilgiris Lions Club, visited. The volunteers included the Canadian CUSO nurse, Meridel

Fowler, two German volunteers with the Deutsche Entwicklungs Dienst (DED), Petra and Christine, and the Australian volunteer, John Holds. The day before this invasion of visitors, I issued clothing soap to all men. All their wrap-around cloths were washed. I had asked them to widen a road to facilitate parking and turning of vehicles for the visitors. Then I bought coloured paper, and Anthony and one of the men made little flags for the children. The guests were expected at 1:00 p.m. I cancelled work at 11:00 a.m. But just before 1:00 p.m., neither the chief nor his brothers were ready! Their bullocks still had to be watered. Another man's bullocks were eating some of the bean plants on our terraces, so I chased them away, in my white shirt and newly polished shoes! The children all held their flags and were waiting eagerly. The idea was to welcome the guests and lead them in with flute and drumming from the corner, one mile away. Under Anthony's directions, an honour gate sprang up at the Colony entrance. Kurian let two of his banana trees be chopped off and installed left and right of the drive into the Colony. A chain with coloured flags was strung high up between these trees, across the entrance. As I tried to hurry the men up, I realized that I wasn't at all exasperated, even though organizing something which wasn't quite ready on time. After all, I wasn't the host. The Paniya Colony was welcoming guests. And why try "Prussian drill" precision anyhow? That would have been too unnatural.

We just got moving a little ways from the Colony towards the road corner, when the Doctor and the first group of cars arrived. I asked the cars to stop and wait for the remaining cars. This gave the flag-waving Paniya children an opportunity to line up on each side of the road. The music started and because of the wait, the Panichis also arrived, and amidst hollering, began their dance. Soon this "procession" moved through the special gate to the parking place prepared in front of the community kitchen. Here we were treated to plenty more music and loads of dancing and shrieking with enjoyment. I did take some photos, this being the first time the Paniyas had performed during the daytime, and I had my camera handy. All morning the women had gone all out in cleaning the outside of their houses, applying fresh cow dung and mud surfacing. It looked really great.

Then the Lions' Club members came and were so much involved in

Dancing Panichis

greeting each other and getting their food out for their picnic that the Paniyas might as well not even have been there! Well, I guess it was obvious that I became really annoyed and disturbed. I hardly ate anything. Anthony, Kurian and the Paniyas were terrific! All so helpful. Anthony was in charge. He helped to carry the food, get the vessels and the water for drinking and for tea. This all took place inside the Paniya community kitchen. The rich, the Lions' Club members, ate their meals while a portable record player played music, while the poor, the Paniyas, who had already had their modest meal in the same location, looked on from the outside, through the slats!

The Doctor and I left at this time to buy some loaves of bread in nearby villages, hamlets and just roadside shops, for each of the Paniya families, to be given to them later, as a treat. We had to try four or five different stores until we had enough bread to distribute. When we returned, I was somewhat appeased as I saw that there was fortunately lots of food remaining from the Club members' meals and this had been given to the Paniyas.

The Doctor gave a speech, in English, to the Club about the Colony and our work here. Subsequent to this, the Lions' Club President promised a sizable financial contribution, the exact amount to be decided at their next meeting. I sure hope it will indeed be sizable! The Doctor is such a good and convincing speaker, with a terrific command of the language, and wonderful choice of words! Later the Lions had their Directors' meeting and at the end of this, at the Doctor's request, a Panchayat (a governing council of five) was elected from among the Paniyas. This included the "chief", Vella, and four of the older men, namely Kaima, Veran, Chundan, and Kavalan. The Lion's Club President wrote down the names of the members, and as he read them out, all other Paniyas lifted their hands, indicating their agreement with this Panchayat. The Doctor went one step further. He asked each of the Paniya men individually to come forward, and to repeat the words that he would obey the decisions and respect the Panchayat members. Later we showed the guests around the Colony. Amused smiles greeted us when the hut I lived in was pointed out to them.

Later one of the German volunteers asked me about the possibility of buying some cloth and having a tailor make clothes for the children. She said this was possible since she felt that their DED salary wasn't really so small, being Rs 600 per month, plus a kerosene refrigerator and other items in kind. Was I ever delighted! And I do hope that this suggestion and offer will materialize faster than the BDO's offer of a gift of clothing. We are still waiting!

Our CUSO-Delhi Coordinator, Bob Pim, cancelled his planned trip to the Nilgiris because of his chronic bronchitis. However, he wrote to me that the Bata Shoe Company, which has many outlets in India, was now considering a gift of shoes, not just for us CUSO Volunteers, but also for our Paniya children. So now I have to sketch the feet of all the Paniya children and provide details for shipping, i.e. by rail from Calcutta. I do hope this too will materialize! It would be so nice and is really necessary.

Imagine my happiness when yesterday I found that eight to ten of the men of our Colony had each bought himself a brand new shirt, from a travelling merchant. I gladly gave them the requested Re 1 wage advance they asked for subsequently, once they showed me the shirts they had bought. A couple of men even bought cute little dresses for their preschool daughters. I'm sure the Doctor will be only too happy, as he always says that they must want something badly enough to save for it! In the recent past the men had definitely told me that they wanted new shirts. Since they bought them for themselves, they obviously wanted them badly enough. Of course, such a shirt won't last very long; it's not of the strongest material. But anyhow! Also, in preparation for the visit of the Lions' Club, quite a few of the men had had their hair cut. And everyone was really quite clean.

Today John and I gathered six of the boys for haircuts, and tea to follow. Five boys remain for next weekend. Again, one of the men came along. Guess what I did this time? I paid for his haircut. John seems like quite a nice chap, quite congenial and agreeable. Since CUSO does like to have all its volunteers come to the annual reunions as often as possible, the Doctor suggested that John "hold the fort" for me here, if I decide to go to the reunion in Delhi in mid-March. I am not quite sure yet if I will go, or indeed will be able to go. Perhaps our tapioca harvest may just start around then. We'll see. Anyhow at present John is staying with me for one week, just for curiosity's sake. Anthony surrendered his cot and is himself sleeping on the ground. John is helping me a lot. He has repaired several door hinges in the Colony homes; together we almost completed replacing my ceiling sacks with nice bamboo mats, since I'll still be in the hut for some months. John also did some sewing of torn children's clothes and helped me copy out the children's foot sizes.

Yesterday a woman became upset over something, and then started a quarrel with another woman. Soon many became involved. Again, the main issue became food and cooking of the midday meal. One group said that another family could move out of the Colony if they so desired. I called the newly elected Panchayat together, for the first time. With Anthony as interpreter and peacemaker, the "warring camps" were soothed. At one point though, one of the Panchayat men took up a stick and wanted to beat the woman, his niece, who had said nasty words to the other woman. Anyhow, this was prevented, and everyone went away in agreement. Was I ever relieved! Especially as several of the upset parties were now smiling and laughing with me.

3 February 1967, Friday

So many things have happened this past while. First there was the visit of the Collector (the top administrator for the Nilgiris District) and his wife to Gudalur. There they distributed gifts from the Welfare Department. Five Paniya men received agricultural implements, the "mumattis" (the hoe/spades), crowbars, and digging forks. In discussions with the District Welfare Officer (DWO), I was able to arrange that two steel plows be allocated to the Paniya Colony as well. Then each of the five Paniyas was to receive 50 kg rice for planting. However, as our Colony lowlands of three quarter acre, could at most use 10 to 15 kg, I requested we (the Paniya Colony) receive only 15 kg rice. John and I attended the function. I had sent the chief, Vella, to receive the gifts for himself and on behalf of the other four men named, as otherwise it would have cost too much to transport. I needed to secure the DWO's permission for this. I am glad he gave it.

After the function, we wanted to rush to catch our bus, which would result in a 2½-mile walk home when we got off. Otherwise, if we missed it, we would have to walk 10 miles home from the nearest bus stop. But the DWO called out: "Mr. Mündel, please join us here." So, John and I joined the table for biscuits and tea with the Collector, his wife, the District Agricultural Officer (DAO), the DWO, plus about half a dozen of the other top civilian officials of the Gudalur Taluk, such as the Block Development Officer (BDO), the Tahsildar (in charge of Taluk taxation and revenue) and others. After briefly looking closely at all the donated implements, we rushed to the bus stop. The bus was ready and about to leave. Since I had previously informed the ticket agent of our coming, he had kindly offered to delay the bus for 5 to 10 minutes, as well as reserve two spaces for us.

Vella was to come with the equipment the following morning, which was Wednesday. He didn't show up. Now it is Friday afternoon, and still no sign of Vella. By noon yesterday, I was quite worried and wanted to send search telegrams to Gudalur. After all, Vella and all the costly equipment for the five men had not shown up! Just then the Doctor's driver, plus Murthy, the assistant to Miss Martin, arrived in the Doctor's jeep to bring me some money and cabbage seedlings that we were in the process of planting. Quickly I asked them to take me in the jeep to Gudalur. John remained at the Colony to look after the cabbage planting. In Gudalur I first went to the place where the distribution had taken place. The BDO greeted me. Since we had the jeep with us, we seized the opportunity and took the long-promised clothes (over three months ago by now) for the Paniya children back with us, plus a man from the BDO's office, who was to hand the clothes over to the school headmaster for official distribution. These clothes were stitched out of cornmeal sacks. This distribution is to be next Monday, as the headmaster is absent today. At last the clothes are here!

The BDO knew nothing of Vella. So we went to the hostel for tribal people. No luck. Although I had previously made arrangements for him to stay there, he hadn't been there. On we went to the Special Revenue Inspector for Tribal Welfare. He wasn't at his home. Then to the Taluk Supply Office, in the hope someone might know where Vella had gone. We saw another jeep just leaving from there with the Taluk Supply Officer in it. He was surprised to see me, but could tell me that Vella did not stay at the hostel but with a barber, who calls himself a social worker, and is well known to the Doctor. This barber has been referred to many times in my presence as "the mad barber of Gudalur". Supposedly for a fee, he tries to represent the tribal people to government officials. He is known for various dealings he has been involved in.

We headed to the barber's house. Vella had apparently left the evening before, but was to return that evening. The barber was not present, only an old woman and all the equipment! For a good half hour we tried to convince her to release the tools to us. This included giving her the licence number of the jeep and a signatory promise by the Gram Sevak[9] (the man who had come along from the BDO's office) that he would be there that evening to explain to the barber what we had done and why. Finally she relented, and we got the equipment. There was speculation in our group that the barber might have paid Vella some cash and that he might have wanted to sell the equipment. Anyhow, we were able to bring the clothes and equipment safely to the Colony. It's actually almost a detective story.

DONATIONS WELCOME BUT HOW BEST TO SEND?

10 February 1967, Friday

I offer an observation on some of the postal money orders relatives and friends have been sending me to assist with the needs at the Paniya Colony: at times the children in the streets tell me that I have a money order waiting for me at the post office. At times they even tell me the amount. This I find *really* unnecessary! The building contractor working on the new house is constantly asking me about money from the Doctor, even if he has not completed the segment to be paid. He told me of a money order waiting for me at the post office, obviously in anticipation that it would contain money to be paid to him! The "joys" of rural and village life! Bank transfers to the private bank account of Miss Armstrong with the National and Grindlays Bank Ltd. should work better. She would be notified by the bank's main branch in Madras and could then withdraw an equivalent

---

[9] "Gram Sevak" is Hindi for Village Level Worker (VLW); from "Gram" meaning village and "sevak" meaning servant. These VLW's maintain village records and are advisors to villagers in matters as disparate as health, education, farming, and general village development.

amount from her local account in Kotagiri to give to me. This seems to me the safest way to receive future contributions for the Paniya Colony. Just last Monday, February 6, I received a letter from Miss Armstrong, confirming that she had been advised by her Bank in Madras that C$50 (Rs 346.74), sent by my parents on January 19, were deposited on February 2. Just two weeks in transit.

I have sent Bata the foot outlines of all the Paniya school children. If I receive this offered gift from them, there are still the preschoolers to think about and some who are too old for primary school. That is where donations would be most welcome.

From donations received from my parents, my grandmother Mündel, and the McHargs (the missionaries I stayed with after Christmas), I paid off many current expenses, and yesterday decided to get each Paniya man a tailor-made shirt. Ready-made shirts are much more expensive here. Today the tailor came and took all the required measurements. Hopefully in a week they'll all be ready, at Rs 6 each (converted that is about 80 cents).

This week I received two Christmas food parcels from home. One was sent October 28th, thus it took 3 months and 12 days. The dark Christmas fruit cake is in excellent condition and very welcome, especially since John is still my guest. A light-coloured cake, sent in the same package, was totally mouldy.

12 February 1967, Sunday

A hunting party from the Moolukurumba tribe, with their bows and arrows and dogs, just wandered by on the road by my hut. Soon I heard loud shouting as they circled around and closed in on their prey. I assume it was a rabbit, but don't really know. Anyhow, they didn't get it. On a recent visit from the Doctor and Meridel, we all went to a nearby Moolukurumban village to treat patients. Then we also went to the Kurumban chief's house. His son strung a bow and handed it to his father, who pulled it and let it twang. Then the chief handed it to me, to test its strength. Or was it to test my strength? I'm really not so sure. It was certainly very taut.

DOUBLE PANIYA WEDDING

13 February 1967, Monday

Time to tell about the double Paniya wedding. This has surely been one of the most interesting events I've experienced since coming here! So much action and so much symbolism! I doubt I will ever be able to find out the meaning behind everything we observed or the background to it. First of all there is the irksome language barrier, and then, I must assume the exact meaning of many old Paniya traditions are unknown, as is common in ancient customs. By the time John, our guide Urali, and I reached the place, some three miles from my hut, it was the evening of the second day of the celebrations, just before dusk. The area around the Paniya hut was full of Paniyas in clean or even new white "mundus" (the cotton cloth the Paniyas commonly wear). It was interesting to see the one bridegroom sitting on one side of the hut, smoking a "beedi" (a tiny wedge-shaped cigarette of tobacco wrapped in a special leaf) and wearing the oldest imaginable looking cloth! No one seemed to be taking any notice of him. Actually, at the particular moment we arrived, quite a bit of the attention shifted to us. A mat was placed on the ground for us to sit on. We just caught the end of some kind of ceremonial walking about by three girls. The two on the outside had small water containers on their heads.

After a while the music started, with the flutist and three drummers all marching in a circle with a whole group of men dancing at times quite wildly around them, their arms often held quite upright. They would jerk their arms around and then down. Much shouting was involved. Meanwhile some 30 feet away, about ten of the women did their characteristic circular dance which I described earlier. After about half an hour of this music and dancing, the two bridegrooms were led into the male circle. With their heads and faces covered by their cloths, the bridegrooms squatted in the centre of the circle. A bag containing 24 small bundles of rice straw was placed beside each squatting bridegroom. The dancing and music continued for a while.

Then male relatives took turns carrying one of the bridegrooms on their left shoulders, each for a short distance. The women now moved their dance closer to this scene, accompanying their dance now with shrill intermittent shrieks. The "mummied" bridegroom, astride his relative's shoulder, was then chanted at by an old man. I think this was the owner of the hut where the wedding was taking place. After some time, they would move some steps forward, and the old man would chant some more at the bridegroom. All around people were shouting as if in approval. Again, the relative and groom moved a few steps. Again, the old man chanted. This chanting relates to wishing the bridegroom a happy married life and the absence of all evil being done to him, and requesting him to look after his new wife well. Then the 24 bundles of straw were placed under the triangular roof of the branch-covered wedding canopy. The elevated bridegroom was carried around this structure three times and then placed onto the straw, where he remained standing until the second bridegroom was in a similar fashion, gradually brought there.

The brides were led from the hut, using part of their old cloths to cover their faces. Their heads were not covered. The brides were placed between the bridegrooms. All music had stopped; everyone gathered around. I have forgotten some things that were done and no longer remember the exact order in which they happened. I think to start with, each of the four young people had his/her hair, face, and arms (for the bridegrooms also the whole upper part of the body), oiled with coconut oil, by an attending relative. The brides are sisters, and as mentioned before, all four young people are from our Colony. Also, all four are still teenagers. One of the brides is only about 14 or 15 years old. The brides' faces now remained uncovered, but for the whole proceedings, the bride and her groom didn't even once look toward each other.

A male relative of both the bride and groom then clasped his hands, folded them, opened them, turned them outwards, repeated this sequence, and then touched the ground. He then took a cloth and folded it like a belt, and facing away from the bride, he slipped it over her head and down to her thighs, where he tied it. This was done to each bride in turn. With male attendants for the bridegrooms and female attendants for the brides, they all in turn had their brand new "mundus" placed on them. Only then was the old mundu, underneath the new one, removed. Each groom also received a wide red cloth belt, which was tied at his side. I think it was now that the four sat down on a bamboo mat which had been placed over the straw bundles. The grooms had earrings put through their ears, while the brides received a series of seven or eight arm rings, some for the upper arms and some for the lower arms. The brides also received rings for their fingers. The one bride received eight rings for each finger of both hands, while the other bride received only a few rings and on only some of her fingers. Bangles and two of the large red-seeded earrings were also placed in the lap of each bride. These earrings were later tied in their cloths. Then each couple received consecrated betel leaf and areca nut to chew. They exchanged the unchewed items and then chewed while looking away from each other. Then the attendants withdrew.

Several of the older men squatted down with bowls in front of them. These were for contributions by the guests for the father of each bride, which in this case was the same man. This took quite a long time as the pieces of money being donated were most ceremoniously placed on the plate with much discussion. The grooms and their brides were then led off into the Paniya hut, accompanied by the brides' father, and a few other people. There, some roasted and ground rice and coconut were consecrated, as temple food. After a while one woman came from the hut and brought some of this temple food on two banana leaves. She and the old man who had done the chanting offered this food to John and me. It was really tasty, even more so than the offering I had received after the funeral "pujahs" (prayers) for Vella's wife. While I assume that everyone inside the hut received some of this special food, no one outside the hut received any, other than John and I.

In the meantime, the meal for the guests was ready. The father of one of the bridegrooms placed a mat for John and me at the side of the wedding canopy, where we sat down beside him. Some of the men from our Colony, who had prepared the food, served us a rice meal with a very interesting and tasty vegetable mixture, not at all overcooked as I have experienced here all too often. While we ate from our banana-leaf "dishes", with our right hands, of course, all the Paniyas around watched and commented on the way we ate. The food had a fair bit of liquid in it, and we certainly weren't experts eating with our hands! Since I did know over half of the people who were attending the wedding, I didn't feel overly self-conscious about eating while being watched. Serving and waiting to eat while the guests eat first is certainly a sign of sincere hospitality. Since I was aware that the Paniyas would not eat until we had finished and left, I decided that we should leave almost as soon as we had finished eating. This was my first Paniya meal and my first Paniya wedding. It was a most thrilling experience!

This past week was the first time I saw fairly large snakes of any kind. Twice I've seen one now. One was a few miles away, and then the day before yesterday I saw one right in front of one of the Paniya houses in the bushes. Both were about six feet long or even a little longer, and quite thick, too. I had a good look at both, even daring to chase one a bit with a stick as it crawled away as quickly as possible. They are not poisonous. I am not sure of their name in English, but suspect they are the 'rat snakes' that I've heard are quite common hereabouts.

I do wish that John could stay here longer, or at least come from time to time to help out if that were possible. I hope the Doctor comes today. There are just so many things to discuss with him. Topics would include: issues related to crops, loans, applications for funds, water rights (with neighbours wanting to take these), and the house being built (which is still unfinished and needs many improvements). There are also the patients needing his attention, and I need money for this week's wages and food for the Paniyas. And yes, we will probably continue to have anemic patients who need special attention. There are quite a few types of anemia. They all concern the body not being able to supply adequate oxygen for the cells, from a lack of hemoglobin, as not enough erythrocytes are produced. Deficiencies in iron, vitamin B_{12} , calcium, cobalt, etc. can all trigger anemias, so the Doctor has explained to me. Thus malnutrition results. While enough food may be available (not in all cases of course), the food ingredients for some reason cannot be properly metabolized.

SANDAL JOURNEY ENDURED; R & R; CUSO BOOK

19 February 1967, Sunday

The Doctor hasn't been able to come for some time now as he has been roped into "electioneering" for the Congress Party. After all, he is well-known and it helps to get votes if he arrives in the various villages and declares his support for the Congress Party. John Holds is still here with me. I am so thankful for the extra pair of helping hands! One of us can accompany the children to school, and when necessary, find some of them and cajole others to come along. A few weeks ago, the parents of one of our Paniya school children started taking turns accompanying, or actually dragging, their "difficult" daughter to school. In some cases they have given her little bribes, like a balloon or 10 nP coins. But anyhow, these are the first Paniya parents who actively try to prevent their school child's truancy.

I remember one of the Panichis coming to the well a few weeks ago, (unpaid), while I was involved in getting the children washed and helping them wash themselves. For the first time, a mother came to wash her school daughter's clothes, using the soap I had provided the children for their clothes. She beat the clothes on rocks placed for that purpose. She also washed her preschool daughter's new dress, presumably bought for her by her or her husband. She then washed the clothes for one of the other smaller school girls.

This week has been hectic, interesting, and all in all, amusing. I was in Ooty to buy sandals for the children. A good part of the money came from a donation from the India Association of Davis, where I had studied for my Masters. Bev had mentioned my needs here to them. Anyhow, in Gudalur I had to wait at the bus stand from 4:00 p.m. to 1:00 a.m., as the bus needed repairing. It was cold so the workmen took frequent rests. I couldn't leave the stand as I had 28 pairs of sandals to look after! So I had fun standing in the people-packed bus stand area playing the "staring" game with curious people. They stared at me and I returned the stares. At various times, a number of people came up to talk to me at some length. Gradually, as it became dark, the crowds thinned out. One boy tried desperately to get me to hire a taxi, so he could come along. One by one the other passengers had lengthy haggling sessions with the ticket agent to get their money refunded. Then they disappeared for the night, to sleep somewhere. Others disappeared into the slightly warmer waiting room and booking office, since it was getting colder and colder as night fell. Finally the last of my conversation partners took shelter and I alone remained outside, guarding the sandals, and still chatting to the Anglo-Indian boy who had wanted me to take a taxi. By now he had given up on that and was telling me about the lies and scams he and his buddies concoct, when the occasion arises, to draw money from "white-skins"! This approach is quite common here, so he told me. They had a few years of schooling; they study at nights, and now have no money to continue. He said all "whites", in fact anybody, would take pity on them. They had to, so he told me. After all, education is so important, especially in a developing country. Soon I was the only one at the bus stand. It pleased me to think that out of necessity, I was the most patient person around. But the bus still wasn't coming! At first I was rather anxious to get to the Colony, as I had the pay for the Paniyas with me. Soon it was too late for that – the Paniyas would not be buying their food that day anyway and I resigned myself to the fact. Now I could afford the long wait, rather than take the taxi. I finally took all my things with me into the booking office to wait with the others, sitting, half-sleeping, like the rest, until the bus finally came. It was quite cold by then. I reached home at 3:30 a.m., walking without moonlight the 2½ miles from the bus stop to our Colony.

Yesterday evening was also "fun". John and I rented bicycles to go 16 miles for some tennis. Mike Congreave, the Assistant Manager of the nearby tea estate, was to join us, coming down on motorbike. Anyhow, halfway there, one of our bikes had a flat tire! We took turns cycling, waiting for the one on foot, handing over the bicycle, and walking. Mike came by when we were about two miles from our destination. I mounted the back seat of the motorbike and guided the one bicycle alongside me, as we drove slowly on. On the way back, at night (after, what turned out to be four punctures to one tire had been repaired at a cycle shop), the chain broke and a link was lost. So, I coasted downhill, on the level parts John pulled me, and uphill we walked. After about one and a half hours of this, Mike came by again, this time going the other way, and we took turns leading the bicycle from the back of the motorbike. We spent the night at Mike's, and next morning I walked and used the chainless bicycle as a child's scooter, all the way to the rental shop. We had to pay just a bit more than for one rental bike. I think John was, on the whole, rather annoyed at these inconveniences; frankly, I thought it was all rather funny. I couldn't help being in good spirits, which, I think, made John even a bit more annoyed. Oh well!

Yesterday I gave the men the new shirts. Two tailors made the 28 shirts in three days! I watched them awhile at work. I was impressed! Long-sleeved shirts, as it gets cool in the early mornings and late evenings, with two pockets each.

By now it's reasonably definite that I will be going to Delhi from the 12th to 18th of March, leaving here about the 7th and returning about the 21st, to attend the CUSO meetings.

24 February 1967, Friday

The Doctor arrived at the Colony for the first time in four weeks. As there were so many things to discuss, and he only had half an hour to spend at the Colony, I went along with him to Kotagiri as well.

26 February 1967, Sunday

Kotagiri

Within about 8 to 10 days I will be heading off to the CUSO reunion in Delhi.

There is so much to consider before I leave. We need to find a contractor to build a well. Ginger needs to be purchased to have ready for planting once the monsoons start. Then I need to arrange things for John Holds, so that he can better look after the Colony while I'm gone.

I've just received a request from CUSO to write an article for a book on CUSO. Some 10 to 15 volunteers have been selected for this from all over the CUSO world, including four of us in India. I have been asked to write about the Paniyas, the Colony, and my work here. The editor (Bill McWhinney) decided to ask me, after having read my report to CUSO in December. He wants a contribution from a CUSO volunteer doing work among the "adivasis". Well – another job![10]

ELECTIONS: INDIRA GANDHI ELECTED AS PM; DMK IN MADRAS; COMMUNISTS IN KERALA; MEETING THE PM

27 February 1967, Monday

Paniya Colony

These days there is excitement everywhere, in all of India, especially here in Madras State and in neighbouring Kerala. The results of the

[10] The CUSO book is *Man Deserves Man: CUSO in Developing Countries*, edited by Bill McWhinney and Dave Godfrey, Ryerson Press, Toronto, 1968; my article is on pages 356-366.

elections were most consequential! While the Congress Party again won the "Lokh Sabha" (the lower house of the Federal government), most of its Ministers were not re-elected. However, Mrs. Indira Gandhi, who became Prime Minister last year on the death of her predecessor, LB Shastri, was elected. Thus she is perhaps the first woman ever to be elected to head a large democratic state.

In Madras State, the Dravida Munnetra Kazhagam (DMK) Party won big. This party has strong anti-Hindi language and anti-northern India leanings. But even so, the DMK did not expect, even in its fondest dreams, such a victory. It now forms the State government! In fact, the leader of the DMK allowed his name to stand for re-election to the "Rajah Sabah" (India's Upper House), where he had served as a Member of Parliament since 1962. And he won his seat there. None of the former Congress Party Madras State Government Ministers held onto their seats.

And Kerala, for a second time in its history, elected a Communist government. The party of leader EMS Namboodiripad, the Communist Party of India (Marxist), forms the State Government in Kerala now. EMS Namboodiripad had been Chief Minister of Kerala for 28 months from the 1957 election, just a decade ago.

A potentially positive aspect of these elections is that now that they are over, some of the officials in the bureaucracy, I hope, will get back to their regular jobs, not electioneering. Nevertheless, partisan politics are involved in most aspects of life here. Most of Dr. Narasimhan's carefully developed contacts, from his involvement in the election campaign for the Congress Party, are no longer in positions of authority. In spite of this, I think it is really quite positive that both Indians and the rest of the world can see that this country is not just a one-party state. The Congress Party is not so powerful that no other party has a chance to affect the governing of the country or the state. An opposition is alive and well, without which the whole idea of democracy would be for naught.

2 March 1967, Thursday

Paniya Colony

Today I really do feel under pressure. John is expected back here tomorrow, and I need to leave so many things ready for him to take over. Since our Colony land has very little grazing left on it, I was able to apply for and obtain a "grazing permit" from the Forest Department. This will permit us to graze our Colony's oxen and water buffaloes on forestry lands. Now I need to ensure that this will work properly, before I head to Delhi. The major purchase of ginger will have to be done after I am gone. I did make inquiries and the necessary negotiations for the ginger, which is to come from a research station in Kerala, via a third party. Planting will begin as soon as the rains start. I also made initial contacts with two potential contractors for the well. We are to receive a partial donation/loan towards the cost of this well from the Sericulture Department (i.e. for silk production) for eventually growing our own mulberries. In order for the loan to remain in effect, a minimum of work on the well must be carried out each week. According to the agreement, construction must begin next week or soon thereafter. We need more paddy (rice) straw for a shelter for our bullocks and also for re-thatching the community kitchen roof. Then too, although the clearing has just begun for the shelter for the bullocks, several of the men need to go almost daily into the forest to get bamboo for its construction, as well as for repairs to the community kitchen. These repairs are, thankfully, almost complete. The various items enumerated above are all outside of any regular routine work done here. But then – that's a laugh too! We really don't have a real routine here! Tomorrow I still need to go to Gudalur to see about sprays, fertilizers and watering cans.

Our CUSO reunion is timed to coincide with the Canadian Centennial celebrations at the Canadian High Commission in New Delhi. And all CUSO volunteers in India are invited to attend.

1 April 1967, Saturday

Back at the Paniya Colony

Since my last entry here, much of my time has been spent away from the Paniyas. Meridel and I travelled to Delhi to attend the CUSO reunion and Canada's Centennial Celebrations. Our trip was an adventure full of interesting stories! I estimate that we covered 2000 miles, from the southernmost point, Cochin, in Kerala, to Delhi. Aside from train and bus travel, we hitchhiked with 23 different individuals, nine of whom had their own drivers. Aside from fascinating scenery, we learnt so much from the people we met, about their area, their specific problems or projects, about the religions of India, and about so many different aspects of life! Our travel expenses to Delhi amounted to C$11 (yes, eleven dollars only). In Delhi, at the Canadian High Commissioner's Residence, Roland Michener hosted an evening with the "Canada Entertains" concert participants (including Tommy Hunter, Gordie Tapp, The Rhythm Pals, Miss Canada, and other performers). It was here that I met, completely unexpectedly, and in my travel-rumpled slacks, Mrs. Indira Gandhi, the Prime Minister of India, introduced to her by our High Commissioner! I greeted her in the manner common here for showing respect, with flat palms of the hands together, the fingers facing upwards. She asked me if I also was a CUSO Volunteer. I replied that that was the case, and that I worked for the Nilgiris Adivasi Welfare Association, with Dr. Narasimhan's group. The next evening was the actual concert, following which the Canadian Press (CP) reporter, who accompanied the entertainers from Canada around the world, asked me for an interview. I was able to tell him about CUSO, NAWA and especially about the Paniya Colony.

During my Delhi stay, I was able to visit the Indian Council of Agricultural Research and purchase a long-needed "Handbook of Agriculture" (covering many aspects of agriculture in India), as well as obtain various relevant agricultural brochures. And the corn breeder gave me 30 pounds of hybrid seed to test. Of course that meant dragging this around with me until I reached the Paniya Colony again!

I also met Dr. Dave Hopper (the Canadian economist working with the Rockefeller Foundation and serving on the CUSO Advisory Board in India), who again told me: "I won't be happy until I get you out of that hole down there!" While I told him that I liked it there (at the Paniya Colony), I understood his concern. I agreed with the other "CUSO-aggies" that in the future agricultural volunteers should not be placed into situations such as I was. As "aggies", they should be placed in positions where their agricultural training could be fully used to positively influence a large number of people. In this connection, I don't really consider myself as 100% "aggie". I don't see things only in terms of efficiency and economics, as perhaps I should! But that is how I am at present.

Today is a special day for Bev and me: one year ago today, in Davis, California, we agreed to become each other's partners for life. Our "secret engagement" remained so for 10 weeks until we could announce it to our respective parents on the same occasion, in person to Bev's parents and by phone to mine. This was on June 10, 1966, the day of my Master's Degree ceremonies. A melancholy time for me now, so far from Bev!

Anyhow, last night John Holds and I were invited over to Mike Congreave's to play cards. John spent the night there to get the early morning bus up to Ooty, so I had a pleasant, peaceful, five-mile walk home in the dark. Well, it was not altogether pitch-black. As I walked along, the stars were twinkling above and a few glowworms and fireflies lit up the sides of the narrow path through the dense jungle brush. With midnight approaching, there was hardly a soul in sight. I was all alone with my thoughts. Really quite wonderful!

As I was writing tonight after dinner, again by the light of my kerosene lamp, four different people sought me out for different purposes. Aside from that, I've been spending my time reflecting on a query of Bev's – namely, whether I felt that I was actually doing something real here. The reality is that I find myself in a situation where things need to be done. I feel quite strongly that my own personal output and contribution are and will remain vastly suboptimal, because of my inadequate and wrongly-slanted academic qualifications. I suppose my greatest frustration, next to the language barrier, is the work itself. I am really so inadequately equipped for the job, or jobs, that really need doing here. Instead of me, someone should be here with a rather thorough training in psychology, sociology, and even anthropology, combined with a much greater dose than I of practical tropical agricultural production and management courses. John Holds would say, "Nonsense, common sense...." But really, common sense, of course, does help a heck of a lot, but the degree of common sense applied to a problem can be increased by adequate training. A much broader base of experience results, even if the information provided in such training stems largely from books and lectures. This information could all be perceived and weighed. I feel that my base just isn't broad enough, not even for the two main areas of the Colony work: the human relationship/comprehension aspect and the agricultural production and management aspect. Both of these are essentials for truly good and effective management of this Colony. I feel inadequately prepared and don't possess the proper tools for doing a good job of teaching agricultural methods and techniques. This fact, together with the great amount of work which needs doing, as well as the daily chores which must be assigned and their execution seen to, result in my being much more of a day-to-day organizer than a more long-range planner and teacher. This is something I can do very little about at the present. But I really must! I do, however, like my work here very much and am ever so thankful for the opportunity of working here.

I plan tomorrow to take as many of the Paniya men as will come, to the barber. This should include especially those who have never had a haircut. I wonder how many of these will come. I can imagine the ladies up the hill (Miss Armstrong and Miss Martins) shouting, "Quick! Quick! Take pictures while they still have long hair. This is history!" I actually think quite a few will want to keep their hair long. Even if not, it is not exactly some great historic moment, to be forever preserved on film. Anyhow, this barber-outing should be fun. I hope.

ON DROUGHT, RAIN, AND A CULTURAL REVOLUTION

5 April 1967, Wednesday

During my absence, some rather unpleasant things occurred. Poor John had to experience this during his tenure! In this severe drought, the beans on the terraces have mostly died. A large portion of the tomatoes and cabbage was eaten by chickens. One of the Paniya families, husband and wife and one child, moved away from the Colony.

But some positive things also occurred. The long-awaited ginger was purchased and is ready for planting once the rains begin. The Community Kitchen roof has been completely re-thatched. The planned cattle shelter is almost complete. The well beside the new house is progressing. And the new house is more or less, for good or bad, completed.

Yes, this new house has major gaps between the roof and the walls in both the kitchen and the bathroom. All manner of insects can thus access the food in the kitchen! The doors only shut if someone applies pressure on the other side. This can be expected to get even worse when the wood swells from the monsoon rains. The windows are only tacked in place with four nails. One window has already fallen out when Anthony and his wife, who now also helps him, tried to wash it.

The Doctor & Co. visited yesterday. Among other things, they brought me funds paid into the Madras account of Miss Armstrong for the Paniyas. The funds came from friends and relatives in Canada and from my grandmother Mündel in Germany. My brother, Hans-Dittmar, and the theological institute in Goettingen, Germany, where he is studying, also contributed to the funds. I plan a shopping trip to Ooty, as very few items are available locally.

8 April 1967, Saturday

About five minutes ago, as I was sitting here and really sweating in the sultry evening, a terrific windstorm broke loose! Hopefully it will bring the long overdue "spring showers!". For some weeks now there have been signs of showers in the distance, clouds coming and again disappearing, and rain in the mountains just 10 miles away. But here – still drought! Drought! Drought! Just now some drops are actually coming down!

About three days ago, during the course of two nights of rampage, a family of three wild elephants wreaked havoc in the banana and areca nut plots only about three quarters of a mile from here. The trees were all broken off. Fortunately no people were hurt!

OK – now it's really raining, pouring in fact! Hurrah! The roof of my hut is leaking something fierce. But this time – ha! ha! – I was faster than the rain, and yesterday had placed my 10 x 10 ft. plastic sheet on the roof over my room. Everywhere else in and around this hut it's wet! Amana (Anthony's wife) is quite worried, food on the open fire and rain dripping into the kettles and onto everything! But I'm thinking that if this keeps up awhile, the soil will soon be soft enough so that we can get going on preparing the land for the ginger. It should be planted around now, but so far the soil has been as hard as rock. And who knows, soon there may be enough moisture in the soil to spare us the boring and time-consuming weekly waterings, by bucket brigade, of the lime and mango seedlings, not to mention the daily watering of the tomatoes, cabbages, and silk cotton. Indeed, this is the first of the spring showers!

Right now I am sitting here on my doorstep with two Paniya men, Karimban K. and Kullavella, who left their firewood chopping to take cover on my veranda. We are listening to music on the short-wave radio which Meridel had temporarily lent me, before we went to Delhi. Thus I've been able to be a bit in touch with the world again. I no longer receive the newspaper. Several of them never did make it to me; that enterprise proved a bit too costly, so I cancelled my

Watering mangoes and limes by bucket brigade

subscription. Now I can listen to broadcasts in the most diverse languages, with news and music from the respective countries: BBC, Malaysia, Japan, China, Hong Kong, Australia, Ceylon, Iran, and some Arab countries, Greece, some French-language station, Voice of America (at certain hours), and, of course, numerous Indian languages. Oh yes, also Russian and the Radio Moscow English Service, and some other Slavic languages.

Also I listened to a number of English language broadcasts from Radio Peking, with a very strong signal and hence reception, at various wave lengths. Each of these broadcasts includes a rant against Lui Shaochi, the President of the People's Republic of China, but he isn't once mentioned by name but as "the top person in authority, within the party, taking the capitalist road", and "you are the Khrushchev lying at our side!". Obviously Radio Peking is pro-Mao, the charismatic Chairman of the Chinese Communist Party, and pro his Cultural Revolution. Liu Shaochi is attacked for putting forward the "decadent bourgeois philosophy of idealism", and being a revisionist, who is "dreaming of restoring capitalism in China, thus once again making people into beasts of burden". Strong attacks were also made on Johnson, Kruschev, Kosygin, and Breshnev. I once listened to three of their "pop" songs, with such titles as: "We Are Proletarian Revolutionaries", "The Great Cultural Revolution", and "Chairman Mao, You're the Red Sun in the Heart of the World's People". Before signing off the English broadcast, some of Mao's sayings are read or someone reads statements from his books. To end the broadcast, the "Internationale" anthem of international socialism is played. A most catchy tune! I think it's quite frightful to watch such a regime from a distance. I can only imagine how much worse it must be to live there under these conditions. I sincerely hope never to have to live under such a mind-crippling regime!

Today I have been at the Paniya Colony half a year.

FIRST BIRTH AT PANIYA COLONY

9 April 1967, Sunday

Just now Anama came from the kitchen where she was working and chatting with her husband, Anthony; she stood in front of me with hands folded, and bade me "Good night, sir". She does this every evening. I do appreciate it.

Today was the first birth in the Colony since my arrival here. And my first-ever experience of even partial midwifery! What could I do to help? That's exactly what I was wondering, just as I was trying to start off by bike, in the middle of the afternoon, to go the post office to pick up a money order for the Paniya wages. (Yes, I have a bike now!) Kaima, one of the older men, the one who first asked if we could help arrange for his youngest son to learn reading and writing, came to me and quite excitedly explained that one of his three daughters -in-law was having great pain in the stomach area. With his graphic gesticulations, I understood that she was in very painful labour. He asked me for some pills. Yes, pills are probably considered cure-alls here. But I was afraid to give any out, not knowing whether they might somehow have a negative effect on the baby or the mother. I accompanied Kaima to where the young woman lay. I was only too willing to help, but didn't have a clue what to do under the circumstances! The poor young woman had been lying and moaning and crying something fierce for several hours already! After watching her for a bit, I realized that just standing there wasn't helping. I also realized that everyone was, in fact, expecting help from me. It occurred to me that theoretically, I could just wait and see, and "sure, give it a try", if the child started to appear but had difficulty in emerging. I realized that I shouldn't just hang around trying to "play the hero", when in fact my time might better be spent trying to locate properly trained people who could really help!

So I changed my post-office-bound course and cycled as fast as I could some five miles to the tea estate hospital, near which Dr. and Mrs. (Dr.) Eappan live. Yes, she was willing to come along, provided the road was reasonable for her car. This I fortunately could assure her of, since just this last week work on repairs had resumed on this road. The biggest ruts were filled in. Slowly, too slowly for me, she got ready. She gave me lemonade, which after my hasty bicycle ride was most welcome. We had to go to Cherambadi to her dispensary, to get and sort her medical kit. This too seemed to take forever. Eventually we arrived at the Colony, and Kaima greeted us with a big smile. His granddaughter had arrived! There in the dark room was the young mother, sitting up, with her first-born on her lap! The young mother was then laid down on some sacks and Dr. Eappan gave her an injection to reduce bleeding, plus instructions for the grandmother on how to look after the little one. Somehow I wondered why the doctor didn't check whether the umbilical cord was tied. Just in glancing, I was quite sure that it was not tied. So I mentioned it. Dr. Eappan said, "Oh yes, they will have tied it!" But alas, it was not tied. Apparently this was the first time in her 35 years of practice that she had come across such a case amongst the tribals, or anybody else, where they hadn't tied the cord right after birth. Anyhow, her lack of thoroughness didn't impress me.

I wonder whether I was maybe seeing this little Paniya child in a state that most "western" mothers, in our sterile hospital wards, never see their infant babies. All bloody and unwashed. As the grandfather said: "A little devil!" But still, such a cute and helpless "devil"! Its little arms shaking and its tiny fingers folded in every which way. I'm not sure if the father, Chadajan, even saw his child at that time. He had been sitting alone in the adjoining room.

Anyhow, this brought home to me once again how useful it would be to have a nurse stationed here so that medical procedures could be done properly and hygienically. Lately there have been a number of patients again. Mumps has been rampant in the Colony. Children, women, men. Whooping cough is still present! On another occasion I took one of the children, with the grandfather, Kaima again, to see Dr. (Mrs.) Eappan, to attend to the child's rectal prolapse. The whole rectum, thoroughly inflamed, was extended a couple of inches beyond the anus. I didn't have a clue what to do, but I am learning.

This evening I started giving English lessons to nine of the Paniya children, after they repeatedly requested them. We'll see if the enthusiasm remains. It was sure there this evening! The children sit in a semicircle around me on my verandah. They repeat simple sentences after me, as a group and individually, while we point appropriately. "This is a boy (girl)"; "It's a boy (girl)"; "This is a house"; "It's a house"; "His (her) name is..." And then they wanted to sing. So we did: "Bah, bah, black sheep, have you any wool? Yes sir, yes sir, three bags full." This song was requested by Subramaniam, who was advanced to the third grade, as he had had some schooling before. He has English classes now too. Then followed several Paniya and Malayalam songs by the children. I have never seen them so utterly happy as when singing these songs, with many verses!

Since I've started a 7:00 a.m. to 11:00 a.m. and 2:00 p.m. to 5:00 p.m. workday here for the Paniyas, I have had to get up at 6:00 a.m. to get things rolling. At our latitude, we have about 11.5 hours of sunlight, with the sun rising these days at 6:30 a.m. and setting at 6:00 p.m. Thus I need to rise in the dark. So, while I try to ensure a three-hour lunch/siesta for the Paniyas to tide them over the hottest part of the day, there is some physical punishment for myself – getting up earlier and not usually getting three hours to myself at noon. We'll see how it goes.

Today I was able to distribute aluminum water/cooking pots to half the Paniya families and iron cooking/frying pans for the other half, purchased with some of the recent donations I received from my family and their friends and associates.

Early tomorrow morning I must get the work started and then rush to the post office, to get the wage money I had planned to get today.

PANIYA WAGES CUT AND SEVERE AUSTERITY; WATER BUFFALOES RETURNED TO PANIYAN; PEACE CORPS AND OTHER VOLUNTEERS

15 April 1967, Saturday

Yesterday, I felt utterly perplexed. I had received a letter from Miss Martin, telling me to increase the daily wages of the Paniya men from Rs 2 to Rs 2.50, but to discontinue the midday meal! So, in fact, they would receive less than before! Big "austerity" program. The finances are so tight! Fortunately the tapioca will be ready to harvest in a month or so. The private donations don't suffice, and the various government grants, with the major changes in government, are not continuing! It looks very bad. I was really depressed yesterday. My throat felt very tight. The Paniyas are so poor! And now, for sure there won't be funds for manure and fertilizer for the crops. Are we thus forced to save the paisas and lose the rupees? The letter went on to instruct me to forthwith discuss every expense with the Doctor - every rope for the bullock yoke, every watering can. This is something I just cannot agree to, as I often don't see the Doctor for weeks, or even for a month. And in this Colony there are just too many unforeseen events requiring expenses to be paid! So, I must just swallow any false pride and ask for funds from whomever and from wherever I can. The agriculture needs are many right now!

Speaking of an unforeseen expense! A week ago, on paying Rs 6, I was able to secure the return of two large water buffaloes belonging to one of the Colony Paniyas, used for the past year by a Chetti landowner, as the Paniyan owed him these Rs 6. When the Paniyan came to tell me that he owned these buffaloes, I was quite taken aback. So, interpreter in tow, I accompanied him the relatively short distance to the house of the Chetti. He agreed that he indeed had the buffaloes and that the Paniyan owed him the Rs 6. I asked if he would return the buffaloes that day, if I gave him the Rs 6. He agreed, but indicated that the Paniyan should return after lunch, as his son was just finishing plowing a field with the buffaloes. So I paid the amount and had the Chetti sign a note indicating that the debt was paid in full. Sure enough, in the afternoon the Paniyan came along past my hut humming the "plowing tune" I had become familiar with. There he was and his two buffaloes! He had agreed to pay me back, by my reducing his wages a bit each week. And this was duly done. Aside from this buffalo story being an example of how the Paniyas are being exploited by others in the area, it was also an indication to me that this Paniyan felt he could really trust me, by bringing it to me to help solve it.

Yesterday I called the Paniyas together to share with them the latest developments. I was truly touched by Anthony and Kurian. They tried to protect me in words, so that the Paniyas would not hold it against me that we now have to operate under a more difficult regime. Anthony and Kurian even told the Paniyas that often I requested all manner of things on their behalf from the Doctor, even if none of them had voiced these wishes or requests.

Our new schedule is not exactly without problems. Instead of starting work at 7:00 a.m., we managed 7:15 a.m. and 7:30 a.m. Then a "siesta-style" break from 11:00 a.m. to 2:00 p.m. Because of the heat it would be great to have a break until 4:00 p.m. and then work again till 7:00 p.m. But after the Paniyas are paid, they must be able to walk to Cherambadi and purchase required supplies. Only when they return home can the cooking of the evening meals begin.

On a recent trip to Ooty, I met a few Peace Corps volunteers who invited me over for dinner. They were very interested in meeting me, as they had hardly met any volunteers from countries other than the US. This group of five girls and three fellows are involved in health and sanitation. Their assignments are for two months at one location, then they move on to another location for two months. I think it is an interesting experiment and definitely fascinating and instructive for the individual volunteer. No danger of boredom! It has its definite disadvantages as well. But so do our two-year CUSO placements, where the loss of perspective may be one such disadvantage. Anyhow, these Peace Corps volunteers seemed very interested in the Paniya Colony.

Yesterday and the day before I had two American volunteer ladies visit the Colony. One was a nurse in Greece and the other a Peace Corps teacher in Afghanistan. They were travelling through India and had heard about our Paniya Colony. Before they departed, one of them placed a Rs 100 note in my hand: "For your large family!"(the Paniyas). So kind! And so much appreciated and needed!

This kind donor will be in San Francisco in two weeks and will look up Bev, my fiancée, to show her pictures and tell her in person about my life among the Paniyas.

FINANCIAL WORRIES CONTINUE; NEW HOUSE SUBCONTRACTING

23 April 1967, Saturday

On a bit of a personal note, I just recently found out that Mike Congreave is emigrating to Australia within a month, and that Kesavan has requested a transfer, "from this desolate place". Thus I will not have anyone for even occasional visits and discussions in English, aside from John Hold's infrequent visits and stays.

Funds are really tight around here! Food prices have risen. I wrote to our CUSO- Delhi Coordinator, Brian Marson, to ask for at least a partial food allowance for the Colony. I now have to include kerosene and items for the house in my NAWA allowance of Rs 25 per week. This is aside from housing expenses and Anthony's salary, which NAWA also pays. I even asked Brian about the possibility of completely subsidizing our food requirements, but only if at the same time NAWA would agree to give the freed-up funds to me to use for the Colony! While I doubt that CUSO will go for this, there is no harm in trying, and the situation really is quite desperate. I do hope some help is forthcoming.

Aside from the financial worries, it is becoming inconceivable to me how I am to run this Colony even halfway effectively. I do make my share of mistakes. There really is too much work for me, so that virtually nothing gets done really properly or efficiently. But now, more and more, I find that my planning for the purchase of equipment and supplies is being thwarted by the lack of understanding of agriculture on the part of the NAWA volunteer treasurer. She holds the purse strings. I feel very strongly that personal problems and power plays must not be permitted to jeopardize the Colony and the Paniyas' welfare. They are so poor. Their livelihood just can't afford additional setbacks! If I thought that somehow I was the one dragging my feet, causing the Paniyas to suffer more, I would have to consider packing my bags and leaving for another posting with CUSO. While realizing that things can go on without me would definitely be a blow to my pride, it may be the best thing. I'd just have to overcome such false pride, as the Paniyas must not be made to suffer.

Another example that there is never a dull moment here has presented itself. The house being built, rather roughly, may be ready for me to move into in about two weeks. Also, the well digging is going on near the house. I have to make sure that the work gets done properly, pay the wages according to what work has been completed, hire a contractor to build the well, obtain cost estimates from sub-contractors for making the bricks, and decide what prices are reasonable to pay for the different jobs. I also have to make special trips to buy baskets, pulleys, and rope for the building of the well. This is all in addition to the regular daily work.

The Colony has another new addition – a little boy born three days ago. In this case too, no doctor was available. I had to tie the umbilical cord and disinfect the area. Quite a hefty little feller here, at six to seven pounds, for such a small-statured people!

The other day I received a rather flattering letter from the Director of the Advisor's Division of Canada's External Aid Office in Ottawa. He had heard about me and my work from Hugh Christie, CUSO's Executive Secretary, (and formerly in charge of the Training Division of Canada's External Aid whom I had met at the CUSO Reunion in Delhi), and also from Bill McWhinney, the co-editor of the CUSO book, and the first full-time Executive Secretary of CUSO. Anyhow, I have been requested to provide information and suggestions concerning agricultural needs in India and where I think Canada might best help. First of all, I'll have to make it clear to him that he was misinformed about my having "considerable experience" in my profession, I don't. Certainly not yet. But I'll try to send some comments or suggestions.

MIDDAY MEALS STOPPED; PANIYAS SEEK OUTSIDE WORK AND PAY; A CHEEKY ENCROACHER; FIREFLY SPECTACLE; END OF SCHOOL YEAR

26 April 1967, Wednesday

It's really rough here these days! In spite of assurances that the Doctor and I could meet weekly, he hasn't been here for 10 days. After two more showers, the land is getting prepared. We need to move on the planting. However, the funds aren't here to purchase the supplies.

Some recent updates are warranted: We no longer provide the midday meals. No money! We have the announced "austerity program" in operation due to lack of funds. On Monday only nine men showed up for work: the others have found day-labour jobs elsewhere. On Tuesday there were 14 men, on Wednesday 13. We are already behind schedule with the land preparation! Meridel will not be joining the Colony as its nurse. She has left for Delhi, looking for other CUSO job opportunities.

29 April 1967, Saturday

Our Delhi Director, Brian Marson, has agreed to up CUSO's monthly contribution for my upkeep from Rs 50 to Rs 75, if I am willing to make a similar contribution out of my annual extra allowance of C$350. This I am willing to do. This is to partly cover my increased expenses related to kerosene, food costs, and house needs, as well as for the Colony. I had recently used C$50 of that allowance for costs associated with the house being built. I moved in yesterday.

Showers occur every couple of days. The work needing to get done is burning under our hands. Special funding allocations, needed for the equipment and supplies required, have not been made. We haven't seen the Doctor for two weeks.

As much as my work at this Colony means to me, and I basically like it, CUSO should never have placed an agricultural volunteer here. A volunteer is too heavy a financial strain on such a poorly funded association as NAWA. Then too, the financial inputs are just not there to be able to make even a partway economically sound and agriculturally efficient unit out of this Paniya Colony. For me, however, the other aspects of this posting are really interesting. But just because I have these ideals and crazy notions, I mustn't expect other agricultural volunteers to want such postings: especially with the realization now, that their effectiveness is greatly reduced because of a lack of financing and a lack of understanding of the agricultural needs within the NAWA administration.

Brian wrote in his letter, which arrived today, "I have a lot of confidence in your ability to meet this new situation." And later, "Keep up the excellent work, and I do hope for the sake of the Paniyas that your dedication quotient is as high as ever." No doubt this is meant to encourage me. However, I don't really feel he knows what I am dealing with here.

Anthony, my trusty cook and general helper, has been really sick with dysentery for close to two weeks now. The last two days he had fever and terrible pains. He vomits everything he tries to eat. He can hardly move about.

Yesterday and today, just two of the Paniya men reported for work. The rest have all gone to cut shrub for a contractor of forestry lands, with higher wages than our NAWA allowances, plus a noon meal. Perhaps this wouldn't have happened if we could have kept up the midday meals. Who knows. Three days ago, all the men reported for work. There was no forestry work that day. So, the Colony is being used as a "stop-gap". And now when so much needs to be done, there are no men to do the work! We can't plow, though the soil is terrific for plowing the planned ginger field. Recently we managed to have three teams of bullocks plow each morning and three teams in the afternoons. It really progressed nicely. Only the iron plow and the two plows with the iron tines were used. No wooden plows anymore. The Paniyas chose this, as they could see how much better the plowing can be done with these new plows.

While I was away in Delhi, one non-Paniyan neighbour built a hut on Colony land, thus becoming a squatter. He even had the audacity to plant plantains into the pits the Paniyas dug, at my suggestion, before Christmas. This is in the area where we had trouble with the grazing encroachers. Kurian and I went to talk to the fellow. He got himself into a really heated rage. I kept unusually calm, even smiling at times. This perhaps made him even more livid. As we left, he threatened Kurian with a beating. It sure can't be easy for Kurian to work for us in NAWA, with low salary and the constant risk of being threatened by encroachers. I really do appreciate Kurian's loyalty in such situations. And I have told him so on a number of occasions. Anyhow, I had to book this hut encroachment with the police as "criminal trespassing". How I detest this police business! But it just has to be done. Otherwise, bit by bit, smaller or larger regions of this Colony land would be lost in this manner.

While I was writing the notes above, a shower passed overhead. And, in my new home, at least half a dozen leaks appeared in the roof over my living/bedroom! Four leaks right onto my bedding! Too much moisture collects on the floor where my mattress lies. I will need a bed frame before the monsoons begin.

Quite often I find myself in situations which at the time are quite serious, but must afterwards in discussion, sound odd to some uninitiated "western" ears. Thus, the other day, Kaima told me that his son's wife has virtually no milk for her two-week-old baby. To make his point to me, he illustrated on himself, in the field full of Paniya men, squeezing and squeezing, but alas, no milk! I had to find a solution. This business of too little or almost no milk is really quite serious. I gave the young mother some special vitamin pills, which will hopefully stimulate her milk output, or even her "letting down" the milk. I also cycled three miles to Dr. Eappan, who suggested mixing local boiled cows milk with three parts of water. We'll see, I hope.

I have recently witnessed some of the most beautiful scenes here. With the onset of the rains, the insect population has jumped a thousand fold. We can hardly keep the ants down. But, it's not only the nuisance and pestering insects that abound now, but also the fireflies. One evening I went out and saw this terrific sight: thousands and thousands upon thousands of tiny lights, all over many of the trees and bushes, blinking off and on, with something of a rhythm on any one tree. Three quick sequences of off-on, followed by a little longer off, of several seconds. Then the sequence is repeated. Really an unbelievable sight – like tiny yellow Christmas lights. Some trees were just full of these blinking "goblins" while others had only a few stray ones surrounding them. There were so many fireflies that, in this pitch-black, cloudless, moonless night, during the off-on sequence, whole areas of the trees were illuminated as during a lightning strike. Meanwhile, other trees were beautifully silhouetted against this blinking background. I wish I could get an entomologist to explain this rhythmic blinking pattern to me, this synchronization of most fireflies surrounding one tree.[11] The light emanates from two of the abdominal segments on the ventral (lower) side of each of these insects. A real "dance of the elves"!

School is over. The children now have a month's vacation. During the last three days of school the Paniya children wrote exams in Tamil. A Tamil teacher had been hired just to teach the Paniyas – in a Malayalam school. But I guess the government considers the tribals their special responsibility. The state language, after all, is Tamil. The exam subjects included arithmetic, drawing, and general knowledge. After each test, the children came to show me their slates on which their marks had been written. I saw some 0's and 1's out of 50! But most did get more. Kaima's youngest son, Vellikan, as well as Subramaniam, got tops in arithmetic; Subramaniam also got tops in Tamil. He is the only one who had previously attended a school; that was in Malayalam, and he also had some English. For some weeks now, Subramaniam has picked one of the biggest roses, almost every day, and handed it to me to place over the picture of my fiancée, hanging on my wall. In spite of the poor performance on the tests, after just three days of Tamil instruction, 15 of our Paniya children were promoted to the next grade.

[11] While writing this book, I checked into this a bit and found as examples that the synchronous flashing of fireflies found in Southeast Asia was described by Hugh M. Smith, a biologist from Washington, D.C., as follows: "Imagine, a tree 35 to 40 feet high thickly covered with small ovate leaves, apparently with a firefly on every leaf, and all the fireflies flashing in perfect unison at the rate of about three times in two seconds." Between flashes the tree was in complete darkness he said (Smith, Hugh M. 1935. Synchronous flashing of fireflies. Science 16:151-152). Over half a century later, in review of a 50-year update of the scientific literature, nine hypotheses are mentioned to explain synchrony (Buck, John. 1988. Synchronous Rhythmic Flashing of Fireflies. II. *The Quarterly Review of Biology*, Vol. 63, No. 3 [Sep., 1988], pp. 265-289).

TAPIOCA HARVEST AND SALE; ANTHONY DIES; FENDING FOR MYSELF AND FEEDING THE SCHOOL CHILDREN

7 May 1967, Sunday

Today was the first real tapioca harvest day! The plants are from 8 to 10 feet tall at harvest. The tubers are very long. To harvest the tapioca, one needs to hold firmly onto the stem and pull the plant out of the soil. The long tubers are then whacked off with a knife. Basically, one or two stems are allowed to reach this stage. In that way, greater storage production takes place in the tubers. During the dry season there was very little growth, but with the beginning of the rains, growth has been prodigious. One man purchased 240 kg (about 500 lbs) from us and plans to buy an equal amount tomorrow. I sure can't complain about boredom! I have been up and about since 6:00 a.m.; the tapioca weighing took place at noon; the sale was completed right there. Hungry and a bit tired, I needed to walk the one mile to my hut and begin to cook my noon meal. Not a very cheering thought! But as Kesavan arrived to check on something related to our second well (which is being dug near the farther tapioca field), we both decided to hike the four miles to his place in Erumad to have lunch there. So far I am saving NAWA some money, by not yet having a new cook and helper.

Yes, Anthony died on May 1, of his dysentery. No doctor was around to attend to him. I will miss him – such a terrific person, and his cheerfulness! Anthony was so much more to me than a "butler". A few days before he died, I sent a telegram to the Doctor, asking him to come. However he was in Kerala. By the time he returned, I had already sent him a telegram informing him of Anthony's death. Dr. (Mrs.) Eappan had suggested I send Anthony to her clinic. But the poor man, after starting out to her clinic, returned after a few minutes telling me: "No, Master; too much paining." Anthony had suffered with this dysentery for almost two weeks. For most of the last week he could neither eat nor keep anything down. For his funeral, we walked the five to six miles behind the simple wooden box, draped in

Kurian with three varieties of tapioca

a black cloth, to the graveyard. There we met Dr. Narasimhan and two of Anthony's five daughters. The other daughters had arrived earlier and were in the funeral procession with us. Everyone threw a handful of soil onto the coffin and remained at the grave side until it was completely covered with soil. Then the grave was decorated with a wooden cross and flowers, as well as burning candles. It was night by now. Thus I took my leave of Anthony.

In a weird way, it's almost fun to have to fend for myself now. But on the other hand, so much just doesn't get done, in the field, with the Paniyas, and around my new house! The week following Anthony's death has been very difficult for me. I feel the ones suffering the most are the Paniya children. I've barely given out any pieces of soap; cleanliness is definitely lacking. Haircuts for the boys are well overdue. English lessons have not been continued for a while. And I rarely even stop to chat with the kids; it's really so awful. However, even though NAWA does not pay for this, I am using my own resources and continue with the breakfast for the Paniya school children, which Anthony used to do. I prepare the rice porridge in a huge pot and fry some small fish to distribute to each child, as they come to my hut before school, bringing their own bowls.

Yesterday was a very rainy day. Perhaps the monsoons are beginning. Great streams developed where the mud was taken out to make bricks for the house, now also for the well. I actually fear for the foundation of the house. Also, several of my window frames have swollen so that they can't be closed any more. Others will no doubt follow! Some of the doors are threatening to do the same. In the evening as it was getting dark, I set about diverting the new "streams" with a mumatti, in an effort to save the foundation. It was so soggy and wet. Soon the rain was pelting me again. I was barefoot. Yes, I now actually do own newly purchased rubber boots, but they are waiting in Ooty for the Doctor to come down to the Colony and bring them along! (Again it has been 2 ½ weeks since the Doctor was last here.) Numerous thorns from the *Mimosa pudica*, "Touch-me-not" penetrated the bare soles of my feet; as well I received numerous scratches from the thorn bushes I had to pull out, as night fell, to make the diversion possible. I quickly planted these bushes – actually just sticking them in the ground – around the almost transparent walls of my new outhouse, to

Paniya school children at breakfast with Henning

provide a living hedge eventually. Filthy, muddy, scratched, and exhausted, I went into my house. Was I ever glad that I still had one of the cans of soup sent to me from friends in California for Christmas! The soup plus a mango was dinner. Both of the past two evenings I actually fell asleep as I was sitting on the bamboo mats on the floor, in one corner of my "dining/reception/medicine" room, after just having finished my evening meal.

As I needed to run errands in Ooty, I stayed at the Selbourne Missionary Home, where I had stayed before, enjoying their hospitality. I was considered a guest, in fact a guest of Canadian missionaries who were staying there for a while, and thus the meals and accommodation were free! Many missionaries come to this hill station during the hot season in the plains. During my 2½ days in Ooty I actually purchased my rubber boots, as well as umbrellas and towels for the Paniya children. I also had a pair of work trousers tailored for me and bought coconut coir mats for my cement floors. My attempts at visiting some of the relevant government officials were largely unsuccessful, and so time-consuming. I spent an afternoon at Kotagiri, having announced this in advance by telegram. Yet still I could not manage to meet up with the Doctor.

At the bus stand in Kotagiri, as I was preparing to return to Ooty, I met a group of musicians waiting for the same bus. They had just played at a church wedding. For my benefit, their leader decided that on the bus to Ooty they would play something. They took out their mandolins, banjos, banjo-mandolins, guitars, and mouth organs and played during the whole 30-mile mountain-road trip. The bus driver enjoyed this break in the monotony as much as any of us. Indian and American folk songs were played. Some of these I "had to" sing. The musicians called all this type of music "jazz". I guess that was in contrast to the church music they played for the wedding, where they all played violins. This was an amateur music group. The leader, who is also the choir leader of the Catholic church, is a forest officer. Another musician is a Revenue Inspector for Tribal Welfare. He somehow knew who I was, and thus was the one who first introduced himself to me. Another musician, the son of a Lutheran Minister, was a mechanic. Thus we went to the Lutheran Church in Ooty the next day.

For the last couple of days we have started planting ginger and bananas. Once this is completed, however, I will have to reduce the Paniya work crew and encourage them to find outside work again. We (NAWA) just don't have the money for wages for everyone! Even the tapioca income will only provide small, intermittent, relief. Our weekly "wage" bill is Rs 375 for a full crew. I am now faced with the job of reducing the work crew while still trying to hold the Colony together, as well as trying to get the urgently needed work done on the land! At the same time I'm trying to settle on reasonable prices for the sale of our tapioca. I am not used to bargaining. Yet here almost everything needs to be bargained! Fixing rates with the well contractors is one example. The contractor for our second well wanted Rs 25 per "cole", a local measure of about 2.25 feet. I ended up paying him Rs 15 per "cole", and adding one rupee for each additional "cole" dug. Even purchases in ordinary, theoretically "fixed-price" shops, can be bargained. Thus I bought blankets for the Paniyas which had been marked Rs 12.50 each, for Rs 9 each. In setting prices for the sale of our tapioca, I am trying to consider the demand. It seems to be quite high, as not many people are harvesting tapioca anymore. The main tapioca harvest season is over. There is, however, competition from rice, which currently is quite abundant in our area. I am trying to get as much as possible for the Colony Paniyas. At the same time I'm trying to be fair, and don't want to make it impossible for poorer people to buy this much desired staple! There are real paradoxes here for me, which can really be nerve-wracking for my conscience.

The well next to the new house is now 28 to 29 feet deep; water has just been hit.

As of the Tamil New Year recently, around May 1, the new state government has changed the official name of this State from Madras State to Tamizhagam,[12] which apparently even Mrs. Gandhi has found difficult to pronounce.

CUSO VOLUNTEERS' VISIT; SCHOOL RESUMES; COMMUNALISM AND CAPITALISM; PANIYAS' WEIGHT LOSS

25 May 1967, Thursday

What a surprise I received Tuesday evening! Suddenly a lorry (truck) from the tea estates drove up and delivered three CUSO volunteers to my new house. It would have been a help if somehow they had announced their coming in advance. Not having a refrigerator, I don't keep extra food around, as it would spoil. They hadn't eaten since breakfast, and when they arrived, it was evening! Well, we managed somehow. But it took awhile! Cookies to stave off the initial hunger helped. Then followed an hour of cooking rice on my little cooker, and another hour to cook dhal. After all, I can only cook one item at a time! A little later, the fellows from the tea estate brought a basket full of vegetables, fruit and some nice cold beer, to help me feed my

[12] In July 1967 Madras State was renamed Tamil Nadu by the Madras State Assembly; this was approved by the Indian Parliament in August 1968.

New house with brick-burning in foreground

guests. What a most welcome gesture! In the meantime, my kitchen is a constant mess. I have to keep reminding my guests that everything must be cleaned of food remains right away, as otherwise it will be ants! ants! ants! But during this visit, aside from my 6:00 a.m. breakfast, I don't worry much about the food. Somehow it gets prepared. And it is fun to have pleasant company for a few days. And since it "never rains, but it pours", yesterday a carload of seven missionaries dropped by for a few hours. They had announced their coming in advance and also that they would bring their own food. On top of that, they brought me not only tins of meat, but also of beans and peas, as well as cookies! I felt as though I were the guest! So, eleven of us, squatting on my bamboo mats in the larger room (medicine/dining room), had lunch. The missionaries ate what they had brought and the four of us CUSO volunteers ate what the two girls had cooked. This included a huge steak, which I had prepared early in the morning, brought to us from the fellow at the tea estate. This was my first meat since moving into this bungalow.

Recently Kurian has been more cooperative than ever. I'm giving him more and more responsibility.

Aside from now encouraging the men to find outside work, I am also making arrangements for the children to attend a tribal residential school. I hope we won't jeopardize the whole idea of this Colony! I am seriously considering, in a few more months, possibly leaving the Colony, at least for a few months. This would be a money-saving measure for NAWA. I am a great financial burden on them. In the past, before I moved into the house, NAWA provided me with a monthly allowance, plus food allowance, the pay for Anthony, and the rent for the mud hut, totaling Rs 225. I feel the Doctor would consider it a black mark against his name if he had to tell me that there just weren't enough funds for me. I'll just have to make that decision. No person is indispensable. And at present I am just too expensive for NAWA.

While my own expenses have risen, my current financial situation allows me more flexibility in using funds for the Paniyas. Since I am drawing on my "emergency fund" from CUSO-Delhi, for my personal expenses, I use NAWA's allowance for me to purchase much-needed fertilizer etc. Of course I keep complete records, as we want to know how much it costs to run this Colony. My own expenses have gone up since Anthony died. I tend to eat more bread and drink instant coffee, and fruit juices. Also I need to pay someone to go to town almost daily to do my purchases, but I don't get any pay for a cook now. My breakfast is my most regularly scheduled meal, around 7:00 a.m. Lunch I tend to eat lately between 1:30 p.m. and 3:00 p.m. However I eat adequately. Often I have fried potatoes with onions and tomatoes for my evening meal, which I prepare in the morning when I prepare the school children's meal. In the evening I then add an egg and reheat the meal. Often I have fruit such as mangoes, bananas, pineapple, jackfruit and papaya. Kurian has given me pineapples from his garden for free. They cost around 40 to 50 paise (6 -7 cents). I really do enjoy the fruit here.

Well over two tons of tapioca have been harvested and sold.

3 June 1967, Saturday

Gudalur (waiting for a bus to the Paniya Colony)

Last weekend I joined the visiting CUSO volunteers, plus John Holds and one of the two German volunteers, Petra Erdmann, for a CUSO Peace Corps get-together in Calicut, Kerala. The night before we left, I had five guests staying at my little bungalow. I slept on a cot in the kitchen. It was too short for me, really. After the Calicut trip, John and Petra came to my place for a day. The two of them spent the whole day cleaning my place up! – cleaning pots, washing floors. Now it does look livable! As I needed to go up the hills anyway, I accompanied John and Petra by bus to Ooty. After all, I had promised the Paniya children the umbrellas by the time that school started again. This was the day before yesterday. All the items which I had purchased weeks ago were waiting at the Selbourne missionary home in Ooty, to be brought down to the Colony in the NAWA jeep. Not wanting to wait any longer, I took the whole load by bus: the umbrellas and towels for the children (one per two children), the blankets for the men, the rubber boots for me, and the rugs. The blankets are needed for the cool mornings and evenings and also as protection against the rain and general moisture.

In preparation for the start of school, I had arranged with a "dhobi" (laundryman) to both clean and mend the children's clothes. I cook "rice-kanji" for the children before school, adding wheat germ, for the vitamins in it. Five more Paniya children will be enrolled in school this term. Alas, no clothes for these new school children yet! I'll have to get a tailor to make some quickly.

For the upcoming growing season we are setting up the farming to practice a combination of communalism and capitalism. Aside from the tapioca, ginger and lemon grass, which we are growing for the whole Paniya Colony, I am providing each man with planting material to plant for himself in areas not yet cropped. Each paise earned from that crop belongs to that Paniya family. The last couple of days I issued 10 kg of ginger to each Paniya family, to plant privately. They have cultivated and prepared their own land for planting this ginger.

It was fun to observe that not only did the Paniya men work longer hours, but they also had their wives and children help. I enjoy seeing this teeming activity. We'll do a similar plan for the tapioca. Up to now we have sold four tons of the past tapioca crop, and in a couple of weeks, more should be ready to be harvested.

Apparently some funds have recently become available to NAWA. I have no idea how much or how long it will last, but I have not been forced to dismiss any of the Paniya men and have them take outside work. I had resisted doing this so far. My suggestion to the CUSO-Delhi Director about absenting myself for some months from the Colony has been quite well received. He is hoping I could get a well-paying job for the time, or indeed for the rest of my stay in India, so that I could help the Paniya Colony financially, as well as visit them from time to time. We'll see what happens.

While I usually get up around 6:00 a.m. nowadays, I often can't get my chores done without interruptions. These sometimes start as early as 6:15 a.m.! I must be strict! No disturbance until 7:00 a.m.! And while some of the children arrive at my house at 6:30 a.m., they have to wait patiently until 8:00 a.m. for their breakfast, as a group. Otherwise I just can't finish getting ready, preparing and eating my own breakfast, preparing the children's breakfast, plus putting something on for my lunch, cleaning my place up a bit, and such.

3 June Cont'd

Back at the Paniya Colony

I'm back now at home. Just this evening I weighed the Paniya men, this being about six weeks after discontinuing the midday meals. At present their typical noon meal may be what they can get in the tea shops: tea, "pan" (consisting of areca nut and anise seed or cardamom, wrapped in betel leaf) and chewing tobacco. Not very nutritious! They have lost on average five pounds each. How terrible! One man lost 12 pounds. I had better concentrate on supplying extra vitamins to those who lost the most weight! Also I am giving each man 5 kg of tapioca gratis, twice a week, depending on availability.

CONSIDERATIONS ABOUT MY DEPARTING; DONATIONS; CHILDREN ENROLLED IN TRIBAL RESIDENTIAL SCHOOL

16 June 1967, Friday

Ever since the financial crisis here, the idea that I must leave has developed slowly, slowly. My nerves are quite wrought. I did finally manage to have a long nighttime talk with the Doctor. He just can't comprehend how I could possibly consider leaving, when I like my work here so much. He was sad. And this made it even harder for me to finally decide that indeed it is best if I do leave. I just finished writing a letter to Brian Marson, our CUSO-India Director, asking him to go ahead with seeing to the formalities and arranging for my transfer. I'll probably leave here early August and go to Phaltan in Maharashtra State.

My objectivity and decisiveness sure haven't been helped by the fact that for the past three weeks I have not at all been well physically. I've had a persistent stomach ache, on and off diarrhea, and on and off feelings of nausea. Luckily I was in Ooty, with the kind McHargs, when the fever struck, with severe headaches, nausea and cold sweats. The McHargs gave me quite a few cans of food, which should make my food preparations easier, for a while. Now I am on two different courses of treatment from two different doctors. Dr. Narasimhan is treating me for bacterial dysentery, though he wonders whether it might not be some form of indigestion. And a Danish lady doctor, whose clinic has a lab attached to it, indicated that my symptoms could indicate ulcers. Dr. N. does not have a lab set up himself. Of course he is, understandably, a bit upset that I went to another doctor. Aside from the laboratory situation, some of these afflictions seem to affect us "Europeans" much more strongly than the Indians, and Dr. N.'s experience is mainly with Indians. What to do? I just swallow my different solutions and pills faithfully and hope the pain will subside. No signs of that yet! We'll have to wait and see. Perhaps it's just psychosomatic!

During my recent visit to Ooty, I finally managed to meet Dr. Geuting, the German Project Leader of the Indo-German Agricultural Program for the Nilgiris, as well as his Indian counterpart, Mr. Ponnaya. Both were very kind and friendly to me and want to visit the Paniya Colony soon. Their efforts now, though, have to be directed to existing farmers and to be able to show quick results. Thus, not the Paniyas! Once more, Dr. Narasimhan's and my efforts to obtain funds and materials for the Colony have failed!

22 June 1967, Tuesday

Paniya Colony

As the children don't have clothes to change into when theirs are dirty or just drenched from the monsoons, I've asked a tailor to come to the Colony to make each child an extra set of clothes. He took all the school children's measurements and I asked him to make a pair of shorts and shirt for each boy and a blouse and skirt for each girl. Using about half of the Rs 346 received from donations at the end of May, from my relatives (both grandmothers, two aunts, and my mother), I selected and purchased the material myself. The tops will be basically yellow and the bottoms partly dark blue. Making one shirt or one pair of shorts costs Re1. The same amount would cover my morning bread for three days.

2 July 1967, Sunday

Selbourne, Ooty

During the past week at the Paniya Colony, I received the transfer certificates for 14 of our Paniya children to attend a nearby Tribal Residential School. And so, the 15 of us marched, umbrellas open, during a pelting monsoon rain, the three miles to the school to enroll the children. Hopefully, within a week, the order to approve an increase in the number will come through so that the remaining eight of our school children can also be admitted. I think I managed to persuade the District Welfare Officer to make this decision himself, and then to inform the Director of Tribal Welfare in Madras, requesting him to ratify the action. Time is of the essence.

Tribal Residential School children and teachers at Kappala

This lousy stomach trouble of mine shows no indication of improving. Instead, every morning I wake up with quite a strong pain which more or less lasts all day. And at odd times I have these crazy feelings of nausea. Two lady nurses here at Selbourne think it might be amoebic dysentery or ulcers. I don't really care what it is, as long as I can get rid of it quickly! It really is such a nuisance. Often it just makes me indisposed to do much of anything. But I can't really afford many days away from the Colony now! So I guess I'll just have to wait until I get to Delhi in August for a thorough medical checkup.[13]

Yesterday was Canada's actual 100th Birthday!

9 July 1967, Sunday

Paniya Colony

Though as a general rule, tribal school children go home only during holidays, I managed to arrange with the headmaster to allow the children to go home every weekend. He agreed that if on Mondays some were missing, he would go with an assistant to round up the truant children. Also, the children's parents can come to the school for visits either before 9:30 a.m. or after 4:00 p.m. Just the other day, as I was bicycling to Erumad, I saw, at a distance, a group of Panichis. On coming closer, I saw 12 of the mothers of our 14 enrolled Paniya children, visiting. I am still quite concerned about the remaining eight children. The "sanction strength" of this school is not enough to accommodate them. The District Welfare Officer (DWO) who was willing to increase the "sanction strength" to permit our Paniya children all to attend this school, has been transferred. I read about this in the newspaper. What now?

I have received a donation of Rs 691 from my eldest brother, Hans-Christoph, a recently graduated medical doctor. Some other donations are still en route. This includes one from St. John's Lutheran Church, in Oliver, BC (the congregation which my parents attend and I attended as a child); one from my best friend, also a recent doctor, Dr. George v. Hahn; and one from St. Mark's Lutheran Church, the congregation which I belonged to in Vancouver, while studying at the University of British Columbia. A few days ago I was able to purchase an ox cart, using the remainder of earlier donations from my relatives and half of a donation from Bill McWhinney, the editor of the CUSO book, *Man Deserves Man*, for which I had been asked to write a chapter.

[13] Actually it took another 30 years, and a 1997 treatment against the *Heliobacterium pylori* which finally cured me of ulcers!

We now use this ox cart, "bullock-cart", almost daily, even in the rain. Just yesterday I had a roof made of bamboo mats, as cover for the cart. I have arranged to use the cart to haul manure that I purchased with my own money from a neighbour to the Colony. From here it goes by head loads to the fields (the Paniyas carry the loads on their heads, in the traditional manner). We use the manure for the mulberry bushes, for the mangoes, for the alfalfa (Luzerne), and some for the tapioca. I was careful not to use the donations from overseas to purchase fertilizer, as I did not want to use these funds where they would just be "swallowed up" by the agriculture here, rather than used for something more specific and concrete. While I considered leasing a cart, this would not have been very useful. A daily lease rate quoted me was Rs 30. In one week of 6 work days, we would have already spent Rs 180. The whole cart cost us just Rs 290!

I have also ordered and made a down payment on the much-needed platform scale. This will make it possible to weigh up to 50 kg at one time. We need this for both the tapioca and eventually for the ginger. Right now we spend hard-to-come-by money renting similar platform scales. Our total tapioca harvest is now around seven tonnes. In a few days, there will be up to a week of harvest. Then we need to wait one or two weeks before more tubers reach optimum size. The lemon grass from the two newest plantings has now been cut. Now I need to find a buyer for the oil. The ginger has sprouted quite nicely and these days the first weeding is being carried out. Then it must be fertilized well and a leaf mulch added. There isn't much fertilizer available in India, quite aside from our financial austerity at NAWA. The Doctor had agreed to a small purchase, but these days much of the fertilizer in the country is being diverted to drought-stricken Bihar State. Fortunately, the monsoons have arrived in Bihar just now. That may ease the shortage a bit. Over the last two days we planted the corn (maize) hybrids which I carried back from Delhi. We arranged these in a "randomized complete block" to be able to evaluate the relative performance of the lines statistically.

"PANIYAN BASHA" EXAMPLES; MELANCHOLY ON PLANNED DEPARTURE

19 July 1967, Wednesday

Over the past few weeks, I have invited the Paniya children, from time to time, to come to my house to look at magazines, draw or copy texts, and to go through my Malayalam texts. I am trying to record some of the differences between formal Malayalam and their Paniya version. Much of everyday language and grammar seems quite different. For example, "cold" in Malayalam (M) is "thanappu", and in the "Paniyan Basha"(PB) is "kuliru"; "door" in M is "vadhile" and in PB is "kapellay"; "small shop" in M is "kadda" and in PB is "pudie"; "what is this?" in M is "id-end-anu?" (in somewhat contracted common usage), and in PB is "ibbi yenna thayi?". Other words seem mostly like dialect or accent versions of Malayalam. For example, in M 'enikkje venam', "I want", would be 'yekke vonum' in PB; and "water" in M is "vellam" which becomes 'vollum' in PB. It creates considerable amusement among the locals to hear me suddenly not using Malayalam sentences, which "everyone" here would understand, and instead using sentences in the "Paniyan Basha" (even though I will leave here in just weeks), which the local Malayalis can generally not understand; at best a few words here or there.

While the melancholy of leaving here soon is very much present with me, I know that I will look back on these last months most fondly. In spite of the recent hardships and heartaches, I will regard this time with great thanks – this time during which I was permitted to live in this little house, all by myself, in the Nilgiri Hills, one of the most beautiful places in the world, on this Paniya Colony Karikkamunda. In my private life I could do and live as I wished, far from the hustle and bustle of the rest of the world. At most, I dealt with the hustle and bustle related to the Paniyas or to the Paniya children

RUMINATIONS ON DISCIPLINE FOR THE CHILDREN AND OWNERSHIP BY THE PANIYAS

21 July 1967, Friday

Paniya Colony - last entries before leaving my job there

One more week to go before the end of my stay at the Paniya Colony! I will go to a full-time job in Phaltan, working in agricultural research, including plant breeding, more in line with what my academic studies have prepared me for than this work with the Paniyas.

The Paniya children can at times be "real devils", or is it I who is a complete zero, when it comes to instilling discipline! This is not something to be proud of, but to recognize. I often ask myself: "Sure, having the Paniya kids well-disciplined would make it much easier for <u>me</u> and would cost <u>me</u> less trouble, especially when I feel down; but, would it necessarily be of so much use <u>to the children themselves</u>?" After all, they are merely happy and cheerful bundles of energy! More so than I am. And it is certainly important that the Paniya children's positive attitudes towards things about them, as expressed by their cheerfulness, not be stifled! That is why I am against too much of this semi-military drill in so many schools here, more marked in the private schools. I am not advocating absence of discipline altogether. I merely think that some self-expression (without every action being channeled by some stranger from outside the family), is important and useful for the child's development. Of course, in schools, some discipline is important and the child has to learn to live and adjust to others about him/her. But I just feel somehow that so much of this school discipline is for the convenience of the administration, for the smooth and efficient running of the daily routines in school, and for saving time and effort, plus the poor teachers' nerves. But is all this really of benefit for the child? I feel some may be, but perhaps most is not. <u>But</u>, I'll <u>definitely</u> have to do more thinking on this matter.

About 10 days ago, one of the Paniya women lost a child through stillbirth. Her body swelled up terribly and her eyes and skin became very pale and yellowish, due to her severe anemia, together with a fever. The Doctor was in Delhi at the time. Since the one Paniya woman had died in January as a result of such severe anemia, I naturally became very worried! I sent a telegram to John at Kotagiri, asking him to come. I know that he can give injections and knows quite a bit about medicine.

Poor John. He was at that time involved with a sick Toda woman who died just an hour after he delivered her to the Ooty General Hospital. This was very tragic. The Doctor had succeeded very well in curing her of TB of the intestines, but then as soon as she felt better, she overate. From a sociological point of view, this is also quite a blow. While Todas generally practice polyandry, this young woman and her husband were very firm in staying only with one another. After three years of marriage, with no child, there was considerable pressure from the tribe on the young man to marry another woman. The two, however, remained firm in wanting just to remain with each other. They thus became staunch revolutionaries in their tribe. But now, even the bereft young husband feels that perhaps this was god's punishment for him, for his not having followed the wishes of his fellow tribesmen.

Because of the panic button I had pressed concerning the sick Paniya woman, John asked Petra to come down here. She did that and treated the sick woman, who was, however, too sick to take back to one of the NAWA hospitals by bus. So we decided to get the Doctor down here as soon as he got back. He came, and some improvement seems to have taken place in the Panichi's health. However, she is still not fit for a long ride up the hills to a hospital.

Before I leave here, there are some additional items I plan to buy with the donations received recently. I am getting small metal trunks for each Paniya family. I will supply these with locks so that they can lock their personal belongings up, and thus be able to collect and purchase a few "extras" without fear of their being stolen by fellow Paniyas. Last December, the Doctor had mentioned his wish that the Paniyas would have lockable trunks in which to store things of their very own – to help them develop a feeling of "ownership". While we in the "affluent west" are so used to an excess of greed, I do find this idea of having to instill a bit of a feeling of greed in the Paniyas very interesting. I also want to purchase more blankets for the women. In addition, the donated funds will provide a small deposit for each man to open a postal savings account, so that saving can in fact be encouraged! Dr. Narasimhan has wanted such savings accounts to help expand the vision of the Paniyas, who have traditionally lived from hand to mouth. The hope is that they can start to save for emergencies and for the future. I keep a detailed record of whose donation, and how much, is used for what purchases. I want to be able to provide each donor with a record of what I received from them and what it was used for. For items I cannot personally purchase, I will leave the funds with NAWA and request that they make the purchases and send me the receipts.

Tomorrow, a group of us is going to the Mudumalai Wildlife Sanctuary, with an elephant ride into the jungle planned for Sunday morning. The group will include Kesavan and his brother-in-law, plus John Holds and Petra Erdmann. This will be the last chance for an elephant ride for me, in the nearby wildlife reserve. I had told myself, even before reaching India, that I'd go on an elephant ride here. Thus, I just had to arrange it, for this, my last weekend at the Paniya Colony. The price of hiring an elephant with mahout for several hours is Rs 10, thus only Rs 2 per person. This is about as much as a first-class seat in a movie theatre.

1968

NILGIRIS ON VISIT

9 May 1968

Ten months after leaving the Paniya Colony, I was able to make my first return visit (May 1968), while on vacation from my second CUSO-job, in Phaltan, Maharashtra State. I stayed with friends at Sultan's Battery in nearby Kerala. On a borrowed bicycle, I rode the

12 miles, passing by the Tribal Residential School where I had enrolled our Paniya children prior to leaving the Colony. From a distance I heard one of the Paniya boys calling: "Kungkalen, Durra vannu! Durra vannu!" (Kungkalen, the Master has come! The Master has come!). The boys greeted me with "Good morning" and soon the others all arrived as well. Good timing. This was the last day of school for the current school year. The school principal arranged for lunch to be served a bit early, to which I was invited. Jackfruit with honey dribbled on it was served by the headmaster, as a special treat for me! Then I could wander the remaining three miles to the Colony with the whole group of Paniya children. Of course, there were always three or four taking turns on the bicycle, which I led.

At the Colony I met the new Manager, Gopal Nair. With him I visited the Paniya men working in the far-flung fields. When we returned to our original starting point two hours later, where we had left the children, they were still all there waiting for me, rather than having gone on for the remaining quarter mile to their homes. I visited all the Paniya homes, and was shown several newborn babies.

The Colony land is now divided in such a way that each family has two acres of land to do with as it pleases, while the remainder is communally farmed. Thus each man works for two days on his own piece of land, and the remaining four days a week, for wages, on the communal land. But even the Paniyas are admitting that they cannot grow too much of their own food crops, as they would end up harvesting each other's crops, in fact, stealing from each other, and in some cases long before the crop has really matured. For example, small tapioca tubers are harvested before reaching optimum size. That reminds me of one day, during my stay at the Paniya Colony, meeting one of the Paniyas along Silver Jubilee Road. He came towards the Colony and had a covered basket on his back. This seemed a bit odd. After greeting him, I asked him what he was carrying. This was in the manner of general interest, I thought. But judging by his reaction, it was a challenge. He started to mutter and actually tremble. Now I became curious. Well, he had helped himself to our communal tapioca. But it was still very tiny. That is what one would have hoped to deter.

In the evening after the wages were paid, I invited the Paniya men to tea and a snack. There were several types of bread or buns available. As the bread cost 16 paise each, they requested that I buy them the less expensive buns for 10 paise each. One of the Paniya men even suggested that I should not spend so much on them, as my funds surely were not unlimited!

The two wells which had been begun during my stay are now completed. A dispensary has been built beside my former house. It has almost been completed except for the roof and external painting. A shed is being built to house the equipment and the communal harvest. Some of the funds for projects, gathered by my relatives, their networks, and friends, have still been making their way here to the Paniya Colony since I left. One example I wish to mention here is the $300.00 raised by the children of Ranch Park Elementary School in Port Coquitlam, B.C.[14] This resulted from my brother Hans-Christoph's contacts and was used to help start a nursery school and kindergarten for the Paniyas. Now the neat little kindergarten/nursery house has replaced the former community kitchen among the three Paniya group houses. Each week Miss Armstrong comes down for a few days, trying to determine how best to organize and run such a Kindergarten. The mothers seem quite interested. Those who are willing to help volunteer with washing the children and their clothes. A young woman is sought to be hired for this Kindergarten, someone who has an open mind and is willing to be guided by Miss Armstrong, the educator. A nurse is desperately needed for the dispensary. Towards this endeavour, I handed Dr. Narasimhan some money which I was able to save during my work in Phaltan. Miss Armstrong also told me that a donation had arrived for this purpose from my parents in Canada.

Paniya children welcoming Henning

[14] After I returned from India, I was able to make a slide-presentation at that Ranch Park Elementary School and answer many of the burning questions the children had.

Kurian, the farm foreman, who reports to the new Manager, recently married and is on his "honeymoon".

I was able to spend several hours with the Doctor, accompanying him on one of his nightly visits of the jungle hospitals.

It was an exhilarating experience!

Henning with two Paniya preschoolers

New kindergarten/nursery surrounded by three Paniya Colony row houses

POSTSCRIPT

A. Still in the 1960s

Between Christmas and New Year's: with Bev

Having married in Lodi, California in September 1968, Bev and I set out for India, and Phaltan again, almost immediately. That Christmas we spent with Hilde Uber, the kind Basel-Mission nurse in Sultan's Battery, in Kerala. From there it was just a short distance to the Paniya Colony. Once again, we passed the Tribal Residential School and had a chance to meet the Paniya children attending there. It was an opportunity for me to introduce Bev to the children she had heard, and even more, read about from my letters of 1966-67. A visit to the Paniya Colony followed, where "everyone" was curious to meet my new wife! At my former house site, building was going on. The dispensary was completed, as was a storage shed. Finally Bev had a chance to see in person this Paniya Colony, and could thus more clearly imagine the life I lead there.

House, dispensary and storage shed

Bev and Paniya school children at Tribal Residential School, Kappala

B. Visits After Some Years

It will be obvious to anyone reading this book that my time with the Paniyas, whatever impact it may have had on the Paniyas, had been an exciting and important time in my life. Naturally I wanted to share some of this, if even by just very brief visits, with members of my growing family. Bev, by then my wife, was able to join me, as mentioned above, a year and a half after my original departure.

Three more visits followed between 1986 and 1991, which was almost a quarter century after my first joining the Paniya Colony in 1966. In 1986 I took along our son on a visit to the Paniyas, during some added vacation time while I was on a safflower research project/ monitoring trip in India for the International Development Research

Centre (of Canada). This was followed in 1989, with a detour, following an International Safflower Conference in Hyderabad, which enabled Bev and me to visit once more. And in 1991, Bev and I, with our two teenage daughters, Stephanie and Veronika, left Islamabad (while I was posted on a secondment to a project on canola development, funded by the Canadian International Development Agency) for a vacation to India, which obviously included a visit to the Paniyas. It was an opportune time for us to leave as it had become "too hot" for us "westerners" living in Islamabad with family, during the first Gulf War.

Here is something of our experiences on these three trips:

January, 1986

With Martin, our son

I knew that Dr. Narasimhan had died in 1972. Of course, there were trepidations about the future of NAWA and the Paniya Colony. However, for the next 25 years, until her death at 88 years of age, in 1997, Victoria Armstrong, took the lead of NAWA and helped it expand to wider, more diverse areas of community support.

It had been 18 years since Bev and I last visited the Paniyas. The most obvious change was the large area of tea plantation extending over the formerly barren hillsides. Tea provides year-round income to the Paniyas. While during the monsoon season, pickings of just the upper leaves, for quality tea, are more frequent, picking is still done during the dry season. Tea plants were provided and a ready market arranged through the United Planters of South India (UPASI).

My former house was almost unrecognizable, with the tropical growth, including flowering vines over the house walls and roof. The whole compound looked very lush. A long house, with indoor toilet, had been added on one side. This provided guest accommodations, where we were able to spend the night, and a large meeting room for the evenings, used also as a teaching facility during the day. A small animal shed had also been added.

Tea harvest on Paniya Colony

House and compound

During our visit, Miss Armstrong arranged for a late-night meeting, which went well beyond midnight, to deal with current Colony issues. She asked me to join in. All the very old and by now older Paniyas remembered me. Some of the young women, my former school girls, even came to chat a bit, in their limited English. It was indeed interesting to see that some of the real "movers and shakers" in this Colony now were some of my former school children! Those whom I enrolled in 1966! From that original group of school children, most girls had five years schooling and the boys had eight. One had graduated from high school. Having basic reading and math skills definitely reduces the risk that the Paniyas, these former bonded-labourers, will be taken too much advantage of in their market and other financial dealings.

Elukan, one of the older former school boys, posed proudly with his wife and group of children.

My 16-year-old son, Martin, was able to hear and record some of the Paniya music.

Elukan and family; Kurian in right rear

Paniya musicians with Henning's son

Some of the Paniyas told me how some of their group had moved away from the Colony and after a year or two moved back. A feeling of belonging was definitely starting to be developed.

January, 1989

With Bev

On this brief visit, Bev was also able to see first-hand the tea plantations, with shade trees and pepper vines growing up them. We stayed at the guest house on the site and were graciously treated by our host the Farm Manager, Mani, who had met me in 1986.

One of the now-very-old Paniya men, Kaima, was eager to show us his private small coffee plantation. It was Kaima, who in 1966 asked about writing and reading for his son, which launched my education endeavour – enrolling Paniya children into school. He, though going quite blind, was still working his land.

Kaima with coffee beans

February, 1991

With Bev and our two daughters, Stephanie and Veronika

As we drove towards the Nilgiri Hills from Kerala we loaded up with ripe pineapples, which we distributed to all the Paniya families at the Colony. It was a cheerful welcome! By then, the Paniyas had met all the members of my family! Our daughters finally saw first-hand, what in the past they had only heard from me in stories or seen at some of our Christmas-time slide shows – images of our family's past, and the various countries we had lived in.

In 1979 the Canadian government established a sponsorship program for refugees fleeing as "boat people" from war-torn southeast Asia. Bev and I helped form a society to sponsor a family. With funds left over after our sponsorship year, we were able to officially change our bylaws to include international development in our mandate. As we had charitable status, we could undertake fundraising and provide Canadian donors receipts for income tax purposes. Thus, with assistance from the Canadian Hunger Foundation, fundraising of our own, and partner donors in Nova Scotia, we were able to channel close to $40,000 towards NAWA and the Paniya Colony.

These were donated for the purpose of adding a dairy to provide milk for the Colony, with some to sell. This is a long-term project in itself, aside from barns needing to be built and cattle to be purchased. People have to be trained to handle dairy cattle and forage plantings, needed to accommodate the year-round feed needs of the cattle. By the time of our visit in 1991, a modest beginning had been made: a multi-cow barn was built and a small group of dairy cattle was purchased.

At that time, we saw quite a few huts, inhabited both by Paniyas and some others, encroaching on the Colony land. But the Paniyas had come back to live in this "home" Colony, for them and their children. A promising development, actually.

Dairy barn at Paniya Colony

Dairy cattle at Paniya Colony

C. NAWA Annual Report Highlights Since My Last Visit

Dear Reader:

You may well be wondering what happened to the Paniyas and the Paniya Colony since my last visit in 1991.While the trials and tribulations of this generational settlement scheme continue, I will share here some of the highlights gleaned from 15 years of NAWA Annual Reports, covering up to 2005.

By now a second generation of Paniyas at this Colony, with a strong sense of community, is showing some ambition to increase the productivity of the land and even to educate its children.

Numerous organizations have been concerned and, in part, supported the work at the Paniya Colony. For example, donations from Oxfam and Quaker Peace and Service in the 1970s provided for the first 10 acres of tea to be planted. Subsequent donations from the United Planters of South India (UPASI) and the Australian Community Aid Abroad helped to increase tea acreage, including some for each of the individual Paniya holdings. A well-maintained tea plantation can be expected to produce for most of a century. By the early 1990s, 12.5 acres of the collective tea plantation, jointly managed by the 25 Paniya families, was providing good income, with profit shares distributed two or three times a year. Eventually this expanded to 30 acres of communally farmed tea and 30 acres for the individual Paniya families. Work was remunerated according to the time put in for the maintenance of the communally farmed tea plantation. At peak periods for plucking leaves and weeding, additional outside Paniya labour is also needed. In recent years a tea nursery was started at the Colony thanks to a grant from Germany, thus reducing considerably the expense of hauling young tea plants. In years when tea prices are low, there is considerable pressure to find additional outside work. This communal and individual tea farming lasted for a decade, from 1984 to 1994. At that time, all the plantations were handed over to the Paniyas themselves. They had become the individual owners of the land and the farm. Thus maintaining the plantings, fertilizing, and harvesting of the tea and the pepper (on the shade trees), and selling the tea to the co-operative tea factory, were all the responsibility of the individual Paniya families.

Educational sponsorships have been added over the years. Now some of these young members of this cooperative venture enter the group discussions and help plan and develop, including making decisions on the distribution of profits – a development that has taken almost 25 years to achieve. More and more women join in the discussion meetings about farm-colony management and affairs.

The small dairy, partly paid for by our Canadian grant, has led to the building of a good cow shed and also a well to provide water for the cattle. Fodder is grown in the wet season. An irrigation system is still needed for the dry season, as is fencing against outside domestic and wild animals. The harder work associated with animal care also demands higher wages.

Medical care continues to be assured by NAWA through its dispensary as well as nutrition and nursery provision for the young children, while the mothers work. A nurse and helper serve at the dispensary. They also visit nearby Paniya and Moolukurumba villages, to help where needed and to advise pregnant and lactating mothers. NAWA's Chief Medical Officer visits at least once a month and advises and treats patients. In general, the Paniyas on the Colony are now better fed and the children healthier than 40 years ago.

With increasing prosperity came better health, more confidence and self-respect, pride and better clothing. However, heavy drinking by some members, especially among the men, lead to laziness and ill health, and increased the burden on working women to meet family expenses for those specific families. With extra cash on hand, something the Paniyas had never experienced before, considerable reckless spending took place.

While the Paniyas are still considered at the bottom of the development ladder, hope remains with the increasingly better-educated young people, in their effort towards self-sufficiency. As was stated in the 2001-2002 Annual Report: "The Paniya Farm has gone beyond the stage of an experiment, as a new way of centering a traditional community, in a rapidly changing environment, emerges."

D. NAWA-founder Updates

Dr. S. Narasimhan received the prestigious Padmashri award. This is a high civilian recognition given by the Government of India for his contributions in medicine and social work.

Victoria Armstrong was granted Indian citizenship, a rare honour for foreigners, for her dedication to tribal welfare, her leadership in providing health, education and economic development.

H-H M.
21, July 2007

FURTHER READINGS ABOUT PANIYAS

Aiyappan, A. 1992 (posthumous). *The Paniyas – an ex-slave Tribe of South India*. Institute of Social Research and Anthropology, Calcutta. 192 pages.

This book is by a renowned anthropologist, who tried to "study" the Paniyas in 1934, but they ran away: thus he had very limited, mainly third party information on them. He tried again in 1972-73, and this time was successful in obtaining Paniya informants for much of the information provided in the book. During the summer of 2007, I was able to read this book, borrowing it through interlibrary loan. The book deals with Paniyas in Kerala, adjoining the Nilgiri Hills Paniyas of present-day Tamil Nadu. The time is within the decade following my stay at the Paniya Colony.

Thurston, Edgar. 1909. (reprinted 1975 by Cosmo Publications, Delhi). "Paniyan." in *Castes and Tribes of Southern India*, edited by Edgar Thurston and Kadamki Rangachari. Madras Government Press. Vol. 6, 57-71.

This was the main reference material I had access to dealing with Paniyas at all, during my preparation to go to the Nilgiri Hills and assume the duties of Farm Manager of the Paniya Colony Karikkamunda, described in this book. That was in the summer of 1966, at the University of British Columbia, where our CUSO Orientation took place during July and August.

http://www.adivasi.net/frames.asp?file=history.html&head=HISTO-RY&left=left-history.htm

This site will bring you the history of the Adivasi Munnetra Sangam Gudalur, an activist group which started ACCORD, an NGO to assist the landless, dispossessed tribal peoples of the Gudalur valley in organizing themselves in their struggles. The group covers such topics as: land struggles, development programs, health, education, economic development and marketing, common community assets, managing institutions. Donations can be made to assist them and the tribal peoples of the Gudalur region, including Paniyas.

ABOUT THE AUTHOR

Hans-Henning Mündel was born on March 31, 1942, in Kosten, then Germany, during World War II, of transplanted German Balts from Riga. By the time he was three and a half years old, Henning had been a refugee twice: first in January, 1945, then again in August, the same year. With his parents and siblings he landed in bomb-devastated Düsseldorf, for the first three years living in a very crowded refugee-aid shelter home. On immigrating to Canada in 1951, the family settled in the fruit-growing southern Okanagan Valley, near Oliver, BC. There the family eventually purchased a fruit orchard and Henning grew up working in the family orchard and helping neighbours near and far in their orchards, during summers and as needed during mornings or evenings when not in school. Henning obtained his first degree, a Bachelor of Science in Agriculture, UBC in Vancouver. Two Masters degrees from the University of California, Davis followed. In June 1966, Henning and Bev Atherstone became engaged.

Henning, 24 years old and a recent Master of Science in Agriculture graduate, joined the Canadian University Service Overseas (CUSO) as a volunteer from Canada to India for three years. Most of the first year was spent as farm manager among the Paniyas of the Nilgiri Hills. That was from 1966 to 1967. The Paniyas had just been moved to that land, donated by the Madras Government to the Nilgiris Adivasi Welfare Association (NAWA), when CUSO made arrangements for Henning to join NAWA. Agriculture was just one of the areas Henning found himself involved in, others being various areas of social welfare, schooling, treating the sick, and all the while living in quite modest circumstances himself.

Almost a quarter century after Henning left the Paniyas, the Nilgiris Adivasi Welfare Association (NAWA) made the following comments in its 1990/91 Annual Report:

> By extraordinary good fortune, Dr. Narasimhan was, at this crucial time, approached by CUSO (Canadian University Service Overseas). He was offered the voluntary services of a young Canadian graduate newly qualified in Agricultural studies. Hans-Henning Muendel joined us in 1966 and laboured unsparingly and wholeheartedly to guide the bewildered Paniyas to clear scrub jungle and to plant tapioca, a quickly maturing crop, which could become a basic food for the colonists. He had qualities of intelligence, initiative, perseverance, endurance and compassion, and to him NAWA and this colony owe a great deal. He was prepared to tolerate harsh living and working conditions, which would be completely unacceptable to even sympathetic paid Indian workers or to present-day volunteers. He welded the 25 families into a working and co-operative group, though handicapped by his lack of knowledge of their language and culture or of South Indian agricultural practices. He was dependent largely on Kurian, who had very limited English, for communication with them. It says much for his personality and impact, that, when he has revisited NAWA's colony, several timers during the intervening 24 years, the original members have greeted him with joy and affection. At this time what this experiment lacked in resources to develop the land was partially compensated by the enthusiasm and determination of Dr. Narasimhan and this young Canadian volunteer. Their efforts were strengthened by their warm and inspiring rapport with the simple Paniyas whom no one had previously respected.

On June 25, 2007, after an initial e-mail contact with NAWA, K.Vijayakumar, in charge of the NAWA administration for most of the past decade, replied to Henning as follows:

> It was indeed a pleasure to hear from you after such a long time. We are also happy that you are planning to bring out a book on the Paniyas with whom you worked during the early years of the evolution of the Paniya Farm. I still remember your … visit years later during which the Paniyas remembered you fondly and with affection. I myself have been working

with NAWA since the later part of 1977 and among all the tribal communities the Paniyas are something special to me because of their simplicity and happy go lucky attitude.

After leaving the Paniyas, Henning joined the Nimbkar Agricultural Research Institute in Phaltan, Maharasthra State, where he stayed two years, initiating a safflower breeding and development project. After the first year there, he returned to California and married Bev. Together they came to India for another year, leaving with a two-week old son, Martin, born in Poona. Henning undertook his Ph.D. studies at the University of Manitoba in Winnipeg, where also their first daughter, Stephanie, was born. A two-year wheat breeding assignment with CIDA through the University of Manitoba was completed in Kenya, where another daughter,Veronika, Bev's and Henning's third child, was born. Three children born on three continents! Germany and Ethiopia followed, with the German Agency for Technical Cooperation (GTZ Ltd.), with Henning as Plant Breeder / Project Manager for starting a Plant Genetic Resources Centre in Addis Ababa. From 1978 to 2007 Henning was a research scientist (Plant Breeder) at the Agriculture and Agri-Food Canada Research Centre, in Lethbridge, Alberta, Canada. Here he was assigned the development of 'new crops', which eventually concentrated on soybeans, safflower and then dry beans. A 2 ½-year secondment, from 1989 to 1991 to a CIDA project in Pakistan was undertaken, with canola development as the objective. In retirement, Henning and Bev (still active as a psychologist) continue to live in Lethbridge, Alberta.

Author contact: hhmuend@shockware.com

For Readers Motivated to Assist this Paniya Colony or NAWA:

1. contact NAWA - Nilgiris Adivasi Welfare Association
   Kota Hall Road, Kotagiri 643-217
   Nilgiris, Tamil Nadu India
   E-mail: nawa@eth.net

NAWA is seeking contributions towards establishing a 10-bed hospital at the Paniya Colony, with facilities for in-patients and an outpatient section. At present the dispensary at the Paniya Colony is only open during the daytime. Patients requiring hospitalization need to be transported to either Sultans Battery or Calicut, in Kerala State, or to the Kolikarai Hospital of NAWA, which is over 150 km from the Paniya Colony, up the Nilgiri Hills. As mentioned in this book, very sick patients cannot safely be subjected to such arduous journeys. NAWA is hopeful that once it starts a basic hospital at the Paniya Colony to serve that region, government grants can be obtained to assist with running costs, such as staff salaries, medicines and special diets.

Donations may be made by a foreign currency check, in favour of NAWA, or by Mail Transfer to NAWA's current account 325, Syndicate Bank, Kotagiri, Nilgiris, Tamil Nadu, India.

Indian donations are eligible for deduction under section 80G of the Income Tax Act of 1961. In this case use NAWA's current account 01000061207, State Bank of India, Kotagiri, Nilgiris.

2. contact NAT - The Nilgiris Adivasi Trust
   The Treasurer of NAT
   c/o Mulji. Crawlboys Farmhouse
   Ludgershall, Wiltshire SP11 9PN
   United Kingdom
   E-mail: nilgirisadivasis@gmail.com

As this is a charity, approved by the British Charity Commission (Registered Number 1095205), funds donated will be increased by 28% as NAT can reclaim tax to that amount.

Carpe Diem Mündel Publishing

ORDER FORM

My order for this book, *My Life Among the Paniyas of the Nilgiri Hills* is as follows:

Printed Book:

____ Copies @ $25.00 $________
(all photos in black and white)

____ Copies of Power Point Show of 40 pictures in book
@ $ 5.00 $________
(includes 38 colour and 2 black and white photos)

NOTE: Power Point only available to those purchasing book and providing a valid e-mail address which will accept a 3 MB file.

SHIPPING AND HANDLING:
Canada (includes GST) @ $ 5.00 per book $________
(will be delivered gratis in the Lethbridge, AB area)
USA @ $ 8.00 per book $________
Overseas surface @ $ 10.00 per book $________
Overseas air @ $ 15.00 per book $________

Electronic Book:

____ eBook(s) (sent by e-mail: approx. 7 MB)
@ $20.00 $________
(includes 38 colour and 2 black and white photos)

____CD of e-book (cost includes mailing within North America)
@ $25.00 $________
(includes 38 colour and 2 black and white photos)
- for MS Windows XP and later versions only

TOTAL ORDER $________

My certified cheque or bank draft or International Money Order in Canadian or US currency (only) made out to "Hans-Henning Mündel" for this total amount is enclosed.

Please ship the book to the following:

Name ____

Mailing Address ______________________________________
(e.g. house and street number)

City, Province/State ___________________________________

Country, Postal / Zip Code ______________________________

E-mail address _______________________________________

Mail this form and payment to:

Hans-Henning Muendel
PO Box 21031, West View Mall
Lethbridge, Alberta
T1K 6X4 Canada

Please notify by e-mail of order form and payment sent:
hhmuend@shockware.com

Note: $2 per book sold plus $3 per Power Point Show sent, will be donated towards the Paniya Colony hospital project of NAWA.

It is anticipated that books will be shipped within two weeks of receipt of payment. Expect notification of mailing by e-mail, if an e-mail address is provided.